SpringerBriefs in Applied Sciences and Technology

Thermal Engineering and Applied Science

Series editor

Francis A. Kulacki, Minneapolis, MN, USA

More information about this series at http://www.springer.com/series/10305

Sandra K. S. Boetcher

Natural Convection from Circular Cylinders

 Springer

Sandra K. S. Boetcher
Department of Mechanical Engineering
Embry-Riddle Aeronautical University
Daytona Beach
FL
USA

ISSN 2193-2530   ISSN 2193-2549   (electronic)
ISBN 978-3-319-08131-1   ISBN 978-3-319-08132-8   (eBook)
DOI 10.1007/978-3-319-08132-8

Library of Congress Control Number: 2014942710

Springer Cham Heidelberg New York Dordrecht London

Printed on acid-free paper

Springer is part of Springer Science+Business Media (www.springer.com)

Preface

A few years ago, I was doing research on thermal plumes emanating from vertical cylinders. In order to validate the numerical model, I wanted to compare the Nusselt number of the cylinder from the simulation to an experimental correlation. I thought finding an experimental Nusselt number correlation for natural convection from vertical cylinders would be a trivial task. Much to my surprise, after consulting many well-known heat transfer textbooks, I found out that this was not the case. Instead, the textbooks presented the criteria for treating the vertical cylinder as a vertical plate, which is valid in cases where the boundary-layer thickness in small compared to the diameter of the cylinder. Unfortunately, if one cannot make this assumption (as it was in my case), they are directed by the authors of the textbooks to some hard-to-find literature. Thus began my journey to search for and compile natural convection Nusselt number correlations from cylinders. This book attempts to be a single, updated source on the subject.

Daytona Beach, FL, May 2014   Sandra K. S. Boetcher

Contents

Nomenclature

A	Coefficient
a	Constant or exponent
B	Coefficient
b	Constant
C	Constant or exponent
C_L	Function
c	Constant
c_p	Specific heat capacity [J/kg-K]
D	Diameter of the cylinder [m]
d	Constant
E	Function
e	Constant
g	Gravitational constant [m/s^2]
Gr_D	Diameter-based Grashof number, $g\beta(T_{cylinder}\text{-}T_\infty)D^3/\nu^2$
Gr_D^*	Modified diameter-based Grashof number, $g\beta qD^4/(k\nu^2)$
Gr_L	Height-based Grashof number, $g\beta(T_{cylinder}\text{-}T_\infty)L^3/\nu^2$
Gr_L^*	Modified height-based Grashof number, $g\beta qL^4/(k\nu^2)$
Gr_r^*	Modified radius-based Grashof number, $g\beta qr^4/(k\nu^2)$
Gr_x	Local Grashof number, $g\beta(T_{cylinder}\text{-}T_\infty)x^3/\nu^2$
h	Local heat transfer coefficient [W/m^2-K]
$\bar{h}$	Average heat transfer coefficient [W/m^2-K]
k	Thermal conductivity [W/m-K]
L	Height of the cylinder [m]
M	Constant
m	Constant, exponent, or function
N	Constant
n	Exponent or function
Nu_D	Diameter-based average Nusselt number, $\bar{h}D/k$
Nu_L	Height-based average Nusselt number, $\bar{h}L/k$
$Nu_{L,fp}$	Height-based average Nusselt number of flat plate

Nu_x	Local Nusselt number, hx/k
$Nu_{x,fp}$	Local Nusselt number of flat plate
Nu_θ	Local Nusselt number of horizontal cylinder hD/k
p	Function
Pr	Prandtl number, $c_p\mu/k$
q	Heat flux [W/m^2]
R	Function
r	Radius [m]
Ra_D	Diameter-based Rayleigh number, $Gr_D Pr$
Ra_D^*	Modified diameter-based Rayleigh number, $Gr_D^* Pr$
Ra_L	Height-based Rayleigh number, $Gr_L Pr$
Ra_x^*	Local modified Rayleigh number, $Gr_x^* Pr$
T	Temperature [K]
$T_{cylinder}$	Temperature at the surface of the cylinder [K]
T_e	Reference temperature [K]
T_m	Average temperature [K]
T_∞	Temperature of the ambient environment [K]
X	Function
x	Function or coordinate [m]

Greek

β	Isobaric coefficient of thermal expansion [K^{-1}]
δ	Boundary layer thickness [m]
θ	Angle of inclination from the vertical (inclined cylinders) or position on horizontal cylinder [degrees]
μ	Dynamic viscosity of the fluid [Pa-s]
ν	Kinematic viscosity of the fluid [m^2/s]
ξ	Curvature parameter, $(4L/D)(Gr_L/4)^{-1/4}$ or $(4x/D)(Gr_x/4)^{-1/4}$
φ	Function
ϕ	Angle of inclination from the horizontal (inclined cylinders) or bisecting angle (horizontal cylinders)

Chapter 1
Introduction

Abstract Natural convection heat transfer from circular cylinders in quiescent environments is a subject which has been studied extensively. Although a well-established topic, to the best knowledge of the author no single source of reference is found for natural convection from circular cylinders for varying orientations (horizontal, vertical, and inclined). This book attempts to be a reference for Nusselt number correlations for circular cylinders of all inclinations including horizontal and vertical.

Keywords Natural convection · Vertical cylinder · Horizontal cylinder · Inclined cylinder · Experimental · Computational · Analytical · Nusselt numbers

Natural convection heat transfer from circular cylinders in quiescent environments is a subject which has been studied extensively. Although a well-established topic, to the best knowledge of the author, no single source of reference is found for natural convection from circular cylinders for varying orientations (horizontal, vertical, and inclined). Moreover, the last major reference often cited is the review paper of Morgan [2], which was published in 1975. This book attempts to be a one-stop resource on the topic.

The most amount of literature exists for horizontal cylinders. Studying the natural convection flow around horizontal cylinders is important for many practical engineering applications as well as benchmarking numerical simulations. At small Rayleigh numbers, the horizontal cylinder behaves like a line heat source. For larger Rayleigh numbers, a boundary layer forms around the cylinder that must be considered. Although many investigations on the subject exist, there is a wide dispersion in analytical, numerical, and experimental data. The dispersion exists in analytical and numerical methods because of using the simplified boundary-layer equations, boundary condition assumptions, and solution domain size. Natural convection experiments are notoriously difficult to perform because it is very hard to isolate the natural convection phenomena from other influencing factors such as heat conduction losses, measurement issues, radiation effects, size of experimental chambers, and external temperature and velocity interferences.

Natural convection heat transfer from circular vertical cylinders is a topic which has also been studied extensively. Although a large amount of literature exists on

© The Author(s) 2014

S.K.S. Boetcher, *Natural Convection from Circular Cylinders*, SpringerBriefs in Thermal Engineering and Applied Science, DOI: 10.1007/978-3-319-08132-8_1

the topic, it is difficult to find correlations in many standard heat transfer textbooks. For the most part, the texts display a criterion for approximating the natural convection heat transfer from a vertical cylinder by using the correlations for a vertical flat plate. This is acceptable if the boundary layer thickness of the vertical cylinder is small compared to the diameter of the cylinder. However, if this is not the case, the reader is directed to hard-to-find literature references.

Like horizontal cylinders, there are many discrepancies in the available data. This is again due to the fact that natural convection experiments are difficult. Also, for analytical and numerical techniques, the boundary layer approximation is often used (all pressure gradients are zero, and streamwise second derivatives are neglected) and the boundary layer thickness at the leading edge is assumed to be zero. The orientation of the cylinder with respect to a surface is also an issue. Many investigations assume the cylinder is floating in space, whereas more realistically, the cylinder is attached to a surface. Lastly, the conditions at the ends of the cylinder, whether they are heated or not, also sometimes cause disparity.

Finally, but not surprisingly, the least amount of work has been done for natural convection heat transfer from round inclined cylinders. The main reason for this is that the natural convection heat transfer from horizontal and vertical cylinders is a two-dimensional problem, and the natural convection from inclined cylinders is a more complex three-dimensional problem. In fact, prior to 1957, no literature on the subject could be found according to Farber and Rennat [1].

References

1. Farber E, Rennat H (1957) Variation of heat transfer coefficient with length-inclined tubes in still air. Ind Eng Chem 49:437–440
2. Morgan V (1975) The overall convective heat transfer from smooth circular cylinders. Adv Heat Transf 11:199–264

Chapter 2
Natural Convection Heat Transfer From Horizontal Cylinders

Abstract A vast amount of literature exists for natural convection from horizontal cylinders. This chapter separates the field into two main time periods: early investigators and modern developments. Early investigators focused on analytical and experimental techniques, and modern investigations have focused on using computational fluid dynamics. Nusselt number correlations are presented from many sources for natural convection from horizontal cylinders.

Keywords Natural convection · Horizontal cylinder · Experimental · Computational · Analytical · Nusselt numbers

2.1 Introduction

Heat transfer from horizontal cylinders has the most amount of literature out of all of the orientations. This is likely due to the symmetric, two-dimensional nature of horizontal cylinders. According to Morgan [51], who in 1975 published an all-encompassing review article on the natural convection from smooth, circular cylinders, there is a wide dispersion in experimental results due to axial heat conduction losses to the supporting structures of the horizontal cylinders, temperature measurement location, interference of the temperature and velocity fields by convective fluid movements, and the utilization of small containing chambers for the experiments. Champagne et al. [7] showed that the temperature is uniform over at least the center third of a heated cylinder if $L/D > 200$, but many early experimental investigators used cylinders having $L/D < 10$.

At small Rayleigh numbers, the heat transfer from a horizontal cylinder behaves like a line heat source. For larger Rayleigh numbers, i.e., $10^4 \leq Ra_D \leq 10^8$, the flow forms a laminar boundary layer around the cylinder [39]. At even higher Rayleigh numbers, it is expected that the flow becomes turbulent.

© The Author(s) 2014
S.K.S. Boetcher, *Natural Convection from Circular Cylinders*, SpringerBriefs in Thermal Engineering and Applied Science, DOI: 10.1007/978-3-319-08132-8_2

3

2.2 Temperature Boundary Conditions

2.2.1 Early Investigators

One of the earliest studies of natural convection from horizontal cylinders was that of Ayrton and Kilgour in 1892 [4]. In the manuscript, the authors looked at the thermal emission of thin, long horizontal wires with 0.031 mm $\leq D \leq$ 0.356 mm and 790 $\leq$ $L/D \leq$ 9000 in air. The heat loss of the wires was calculated by multiplying the current by the potential difference in the wire. Morgan [51] correlated the experimental results into the following equation.

$$\mathrm{Nu}_D = 1.61(\mathrm{Gr}_D\mathrm{Pr})^{0.141} \tag{2.1}$$

for $10^{-4} \leq \mathrm{Gr}_D\mathrm{Pr} \leq 3 \times 10^{-2}$.

In 1898 and 1901, Petavel [56, 57] performed experiments on a thin wire with a diameter of 1.1 mm and aspect ratio of $L/D = 403$ in air. Morgan [51] correlated the data

$$\mathrm{Nu}_D = 1.05(\mathrm{Gr}_D\mathrm{Pr})^{0.14} \tag{2.2}$$

for $0.1 \leq \mathrm{Gr}_D\mathrm{Pr} \leq 3 \times 10^2$, and

$$\mathrm{Nu}_D = 0.562(\mathrm{Gr}_D\mathrm{Pr})^{0.25} \tag{2.3}$$

for $3 \times 10^2 \leq \mathrm{Gr}_D\mathrm{Pr} \leq 2 \times 10^5$.

Kennelly et al. [35] studied the free convection from small copper wires in air. The diameter of the wire studied was 26.2 mm, and the aspect ratio was 4540. The correlation of data [51] from that paper results in the following for $10^{-2} \leq \mathrm{Gr}_D\mathrm{Pr} \leq$ 0.3.

$$\mathrm{Nu}_D = 0.945(\mathrm{Gr}_D\mathrm{Pr})^{0.118} \tag{2.4}$$

In 1911, Wamsler [69] experimentally determined natural convection from a horizontal thin cylinder in air for 20.5 mm $\leq D \leq$ 89 mm and 34 $\leq L/D \leq$ 147. The data resulted in the following Nusselt number correlation [51]

$$\mathrm{Nu}_D = 0.480(\mathrm{Gr}_D\mathrm{Pr})^{0.25} \tag{2.5}$$

for $3 \times 10^4 \leq \mathrm{Gr}_D\mathrm{Pr} \leq 3.5 \times 10^6$.

Langmuir [42] performed experiments on horizontal platinum wires in air with diameters between 0.004 cm and 0.0510 cm. Morgan [51] correlated the results

$$\mathrm{Nu}_D = 0.81(\mathrm{Gr}_D\mathrm{Pr})^{0.065} \tag{2.6}$$

for $4.5 \times 10^{-5} \leq \mathrm{Gr}_D\mathrm{Pr} \leq 10^{-2}$, and

$$\text{Nu}_D = 1.12(\text{Gr}_D\text{Pr})^{0.125} \tag{2.7}$$

for $10^{-2} \leq \text{Gr}_D\text{Pr} \leq 0.6$.

In 1922, Davis [14] performed natural convection experiments on thin horizontal wires ($D = 0.0083$ and 0.0155 cm) in several fluids: air, CCl_4, aniline, olive oil, and glycerin. The following correlation for $10^{-4} \leq \text{Gr}_D\text{Pr} \leq 10^6$ is

$$\text{Nu}_D = 0.47(\text{Gr}_D\text{Pr})^{0.25} \tag{2.8}$$

In 1924, Rice [61] experimentally determined the Nusselt numbers from horizontal cylinders in air. The data correlate into the following equation

$$\text{Nu}_D = \frac{2}{\ln(1 + 2/0.47(\text{Gr}_D\text{Pr})^{1/4})} \tag{2.9}$$

for $10^{-2} \leq \text{Gr}_D\text{Pr} \leq 10^4$. And in 1923 [60], Rice experimentally determined (and Morgan [51] correlated) that

$$\text{Nu}_D = 0.97(\text{Gr}_D\text{Pr})^{0.203} \tag{2.10}$$

for $4 \times 10^3 \leq \text{Gr}_D\text{Pr} \leq 6 \times 10^6$.

Also in 1924, Nelson [53] studied the natural convection of hot wires in various liquids. The experimental study was that of a horizontal cylinder that was 0.033 cm in diameter and 16.9 cm long. Heat loss results were obtained for water and alcohol for $1.4 \leq \text{Gr}_D\text{Pr} \leq 66$ and correlated by Morgan [51]

$$\text{Nu}_D = 1.32(\text{Gr}_D\text{Pr})^{0.102} \tag{2.11}$$

Koch [37] performed experiments in air in 1927. He used cylinders with aspect ratios between 20 and 152. The correlated results [51] are as follows:
For $4 \times 10^3 \leq \text{Gr}_D\text{Pr} \leq 4 \times 10^5$

$$\text{Nu}_D = 0.412(\text{Gr}_D\text{Pr})^{0.25} \tag{2.12}$$

and for $4 \times 10^5 \leq \text{Gr}_D\text{Pr} \leq 6 \times 10^6$

$$\text{Nu}_D = 0.286(\text{Gr}_D\text{Pr})^{0.28} \tag{2.13}$$

In 1929, Nusselt [54] analytically determined heat transfer coefficients for horizontal cylinders in air or liquids for $10^4 \leq \text{Gr}_D\text{Pr} \leq 10^8$.

$$\text{Nu}_D = 0.502(\text{Gr}_D\text{Pr})^{0.25} \tag{2.14}$$

Schurig and Frick [64], in 1930, experimentally determined average Nusselt numbers for horizontal bare conductors in air for $45 \leq L/D \leq 286$ and $2.7 \times 10^3 \leq \mathrm{Gr}_D\mathrm{Pr} \leq 8.2 \times 10^5$. Their findings were correlated by Morgan [51] and appear below.

$$\mathrm{Nu}_D = 0.57(\mathrm{Gr}_D\mathrm{Pr})^{0.24} \tag{2.15}$$

In 1932, Ackermann [1] experimentally determined natural convection heat loss from a horizontal cylinder with an aspect ratio of 2.8 in water for $10^7 \leq \mathrm{Gr}_D\mathrm{Pr} \leq 4.5 \times 10^8$. The following correlation [51] resulted.

$$\mathrm{Nu}_D = 0.14(\mathrm{Gr}_D\mathrm{Pr})^{0.32} \tag{2.16}$$

Also in 1932, King [36] analytically determined Nusselt numbers for horizontal cylinders in air and/or liquids.
For $10^3 \leq \mathrm{Gr}_D\mathrm{Pr} \leq 10^6$

$$\mathrm{Nu}_D = 0.53(\mathrm{Gr}_D\mathrm{Pr})^{0.25} \tag{2.17}$$

and for $10^6 \leq \mathrm{Gr}_D\mathrm{Pr} \leq 10^{12}$

$$\mathrm{Nu}_D = 0.13(\mathrm{Gr}_D\mathrm{Pr})^{0.33} \tag{2.18}$$

In 1933, Jodlbauer [33] performed experiments in air with $55 \leq \mathrm{AR} \leq 140$ and $3.9 \times 10^4 \leq \mathrm{Gr}_D\mathrm{Pr} \leq 3.6 \times 10^6$. Morgan [51] correlated the results

$$\mathrm{Nu}_D = 0.480(\mathrm{Gr}_D\mathrm{Pr})^{0.25} \tag{2.19}$$

Jakob and Linke [32], in 1935, performed experiments in liquids for an aspect ratio of 4.3.
For $10^4 \leq \mathrm{Gr}_D\mathrm{Pr} \leq 10^8$

$$\mathrm{Nu}_D = 0.555(\mathrm{Gr}_D\mathrm{Pr})^{0.25} \tag{2.20}$$

and for $10^8 \leq \mathrm{Gr}_D\mathrm{Pr} \leq 10^{12}$

$$\mathrm{Nu}_D = 0.129(\mathrm{Gr}_D\mathrm{Pr})^{0.333} \tag{2.21}$$

In 1936, Hermann [31] analytically determined the average Nusselt numbers for horizontal cylinders in gas environments. The following equation is valid for $10^4 \leq \mathrm{Gr}_D\mathrm{Pr} \leq 5 \times 10^8$.

$$\mathrm{Nu}_D = 0.424(\mathrm{Gr}_D\mathrm{Pr})^{0.25} \tag{2.22}$$

In 1938, Martinelli and Boelter [47] (along with a similar study in 1940 [48]) studied the effect of vibration on the heat transfer by natural convection from a horizontal cylinder. The investigators placed a heated cylinder that was 19.05 mm (0.75 in.) and

320.675 mm (12.625 in.) long. The investigators found a critical Reynolds number, based on the angular velocity of the vibrations and the ratio of the displacement amplitude to the tube diameter, for which the effects of the vibrations increased the rate of heat transfer. The authors caution the reader about the results because the work revealed that the diameter–displacement amplitude ratio was a descriptive variable; however, the effect of this variable was not discernible in the experiments.

In 1942, Lander [41] presented Nusselt number curves for natural convection from horizontal cylinders for gases and liquids. The following correlation is for $10^3 \leq Gr_D Pr \leq 10^7$

$$Nu_D = 0.49(Gr_D Pr)^{0.25} \tag{2.23}$$

and for $10^8 \leq Gr_D Pr \leq 10^9$

$$Nu_D = 0.12(Gr_D Pr)^{0.33} \tag{2.24}$$

In 1948, Elenbaas [18] determined that for horizontal cylinders

$$Nu_D \exp(-6/Nu_D) = \frac{Gr_D Pr}{235 f(Gr_D Pr)} \tag{2.25}$$

where $f(Gr_D Pr)$ has been determined experimentally. For $Gr_D Pr < 10^4$, $f(Gr_D Pr) = 1$.

Senftleben [65] in 1951 for air, gases, and liquids determined the average Nusselt numbers for horizontal cylinders. For $10^5 \leq Gr_D Pr \leq 10^8$

$$Nu_D = \frac{2}{X}\left[1 - \frac{0.033}{X(Gr_D Pr)^{0.25}}\left\{\left(1 + \frac{X(Gr_D Pr)^{0.25}}{0.033}\right)^{0.5} - 1\right\}\right] \tag{2.26}$$

where

$$X = \ln\left[1 + \frac{4.5}{(Gr_D Pr)^{0.25}}\right] \tag{2.27}$$

and for *large* values of $Gr_D Pr$

$$Nu_D = 0.41(Gr_D Pr)^{0.25} \tag{2.28}$$

In 1953, Kyte et al. [40] investigated the effect of reduced pressure (0.1 mmHg to atmospheric) on the natural convection of horizontal cylinders. At reduced pressures, the thickness of the boundary layer becomes large, the gas becomes rarefied, and free-molecule conduction becomes important. Kyte et al. experimentally determined the average Nusselt numbers for a cylinder of diameter 0.078 mm and aspect ratio of 1910. For $10^{-7} \leq Gr_D Pr \leq 10^{1.5}$

$$Nu_D = \frac{2}{\ln\left[1 + 7.09/(Gr_D Pr)^{0.37}\right]} \tag{2.29}$$

and for $10^{1.5} \leq Gr_D Pr \leq 10^9$

$$Nu_D = \frac{2}{\ln\left[1 + 5.01/(Gr_D Pr)^{0.26}\right]} \tag{2.30}$$

In 1954, Collis and Willams [13] performed experiments on horizontal platinum wires with 0.0003 cm $\leq D \leq 0.0041$ cm in air. In order to establish negligible end effects, all wires had an aspect ratio $L/D \geq 20,000$. For $10^{-10} \leq Gr_D Pr \leq 10^{-3}$, Morgan [51] correlated the results

$$Nu_D = 0.675(Gr_D Pr)^{0.058} \tag{2.31}$$

Etemad [19], in 1955, performed natural convection experiments on rotating horizontal cylinders in air. The cylinders had diameters of 60.4 and 63.5 mm, and aspect ratios of 7.1 and 7.5, respectively. Etemad tested both rotating cylinders and stationary cylinders. For the stationary cylinders, Etemad found that for $1.2 \times 10^5 \leq Gr_D Pr \leq 1.3 \times 10^6$, the average Nusselt number is

$$Nu_D = 0.456(Gr_D Pr)^{0.25} \tag{2.32}$$

In 1955, another investigator who studied the effect of vibration on natural convection heat transfer was Lemlich [43]. Lemlich experimentally studied heated wires with a diameter of 1.01 mm and aspect ratio of 917 in air. For the case with no vibration and $6 \times 10^2 \leq Gr_D Pr \leq 6 \times 10^3$, Morgan [51] correlated the experimental data into the following equation

$$Nu_D = 0.45(Gr_D Pr)^{0.22} \tag{2.33}$$

In 1956, van Der Hegge Zijnen [68] combined correlation equations found in the literature into a single correlation equation with a wider range of applicability. He reviewed the current literature and took the empirical correlations of Rice [61] (Eq. 2.9), Elenbaas [18] (Eq. 2.25), and Senftleben [65] (Eq. 2.26) and combined them into a single correlation shown below.

$$Nu_D = 0.35 + 0.25(Gr_D Pr)^{1/8} + 0.45(Gr_D Pr)^{1/4} \tag{2.34}$$

For low Grashof number, in 1956, Fischer and Dosch [25] experimentally determined average Nusselt numbers in air for horizontal cylinders with diameters of

0.0022–0.114 mm and aspect ratios between 807 and 12800. For $3 \times 10^{-5} \leq Gr_D Pr \leq 8 \times 10^{-3}$. Morgan [51] developed the following correlation

$$Nu_D = 0.862(Gr_D Pr)^{0.0678} \tag{2.35}$$

In 1956, Beckers et al. [6] performed natural convection experiments on very thin horizontal platinum wires (0.025–0.66 mm) in air, alcohol, paraffin oil, and water at very low Grashof numbers $10^{-8} \leq Gr_D \leq 1$. They concluded that at very low Grashof numbers, the Nusselt number is not dependent on the Prandtl number. The correlation given in [6] is

$$Nu_D = 0.95 \, Gr_D^{0.08} \tag{2.36}$$

In 1958, Kays and Bjorklund [34] experimentally determined heat transfer coefficients for rotating horizontal cylinders. As part of their study, they tested nonrotating cylinders. Their experimental data verified the correlations presented by Etemad [19], Eq. (2.32), and McAdams [49].

In 1960, Tsubouchi and Sato [66] experimentally determined natural convection coefficients for horizontal wires in air. The range of diameters used in the study was from 0.00489 to 0.061 mm and anywhere from 40 to 150 mm long depending on the wire diameter. For $10^{-8} \leq Gr_D \leq 10^{-1}$

$$Nu_D = 0.812 \, Gr_D^{1/15} \tag{2.37}$$

In 1961, Fand and Kaye [20] investigated the effect of acoustic vibrations on the natural convection from horizontal cylinders in air. The test specimen was a cylinder that was 19 mm in diameter and 159 mm long. For a baseline case, Fand and Kaye determined the heat transfer from the cylinder without sound. These data were correlated by Morgan [51] and presented here for $10^4 \leq Gr_D Pr \leq 4 \times 10^4$

$$Nu_D = 0.485(Gr_D Pr)^{0.25} \tag{2.38}$$

Rebrov [59], in 1961, performed natural convection experiments on horizontal cylinders with $1.31 \text{ mm} \leq D \leq 9.9 \text{ mm}$ and $61 \leq L/D \leq 458$ in rarefied air for $10^{-7} \leq Gr_D Pr \leq 4 \times 10^8$. The experimental results were correlated into the following equation

$$Nu_D = \left[0.98 - 0.01(\log Gr_D Pr)^2 \right] (Gr_D Pr)^x \tag{2.39}$$

where x is

$$x = 0.14 + 0.015 \log(Gr_D Pr) \tag{2.40}$$

Also in 1961, Zhukauskas et al. [73] studied the effect of ultrasonic waves on the heat transfer of bodies in water and oils. The diameter of the horizontal cylinder was 8 mm, and the aspect ratio was approximately 20. For the baseline case without ultrasound, the following average Nusselt number expression was obtained.

$$\mathrm{Nu}_D = 0.50(\mathrm{Gr}_D\mathrm{Pr})^{0.25} \qquad (2.41)$$

The preceding equation is good for $1.5 \times 10^4 \le \mathrm{Gr}_D\mathrm{Pr} \le 2.5 \times 10^6$.

In 1962, Deaver et al. [15] experimentally investigated the effect of oscillations on a horizontal wire of 0.178 mm diameter and approximately 2 m long in water. For the case without the vibrations, the average Nusselt number was found to be

$$\mathrm{Nu}_D = 1.15(\mathrm{Gr}_D\mathrm{Pr})^{0.15} \qquad (2.42)$$

for $0.2 \le \mathrm{Gr}_D\mathrm{Pr} \le 20$.

In 1963, Fand and Kaye [21] studied the effect of vertical vibrations on the heat transfer of a 22.2-mm-diameter cylinder in air with an aspect ratio of approximately 25. Morgan [51] correlated the data for the control case (no vibration) which resulted in the following equation

$$\mathrm{Nu}_D = 0.495(\mathrm{Gr}_D\mathrm{Pr})^{0.25} \qquad (2.43)$$

for $2 \times 10^4 \le \mathrm{Gr}_D\mathrm{Pr} \le 6 \times 10^4$.

Another study on the effect of vibration on the natural convection from horizontal cylinders was done by Lemlich and Rao [44] in 1965. Lemlich and Rao tested a cylinder with $D = 1.25$ mm and $L = 490$ mm in water and glycerin. For the case with no vibration, Morgan [51] correlated the data. For $1.8 \times 10^2 \le \mathrm{Gr}_D\mathrm{Pr} \le 1.9 \times 10^3$

$$\mathrm{Nu}_D = 0.58(\mathrm{Gr}_D\mathrm{Pr})^{0.25} \qquad (2.44)$$

In 1966, Tsubouchi and Masuda [67] performed natural convection experiments in air on horizontal cylinders which were 21.5 mm in diameter and 200 mm in length. The purpose of the study was to compare cylinders with grooved surfaces with cylinders of smooth surfaces. For the smooth-surfaced cylinders, Tsubouchi and Masuda developed the following formula for $2.3 \times 10^4 \le \mathrm{Gr}_D \le 7.5 \times 10^4$

$$\mathrm{Nu}_D = 0.44\, Gr_D^{0.25} \qquad (2.45)$$

Also in 1966, Penney and Jefferson [55] performed experiments to investigate the heat transfer from an oscillating horizontal wire to water and ethylene glycol. The diameter of the wire was 0.20 mm, and the aspect ratio of the wire was 752. As a baseline check, Penney and Jefferson obtained data for free convection (no oscillations) only. The experimental data were correlated by Morgan [51] for $0.25 \le \mathrm{Gr}_D\mathrm{Pr} \le 30$ and is displayed below

$$Nu_D = 1.08(Gr_D Pr)^{0.213} \tag{2.46}$$

In 1967, Saville and Churchill [63] did a theoretical and analytical study of laminar free convection in boundary layers of horizontal cylinders. They found that for $Pr = 0.7$

$$Nu_D = 0.548 \, Ra_D^{1/4} \tag{2.47}$$

Saville and Churchill also investigated the limits for which Pr approaches 0 and infinity [12]. For $Pr \longrightarrow 0$

$$Nu_D = 0.599 \, Ra_D^{1/4} \tag{2.48}$$

and for $Pr \longrightarrow \infty$

$$Nu_D = 0.518 \, Ra_D^{1/4} \tag{2.49}$$

Also in 1967, Mabuchi and Tanaka [46] studied the effect of vibration on natural convection on horizontal wires. The fluids tested were air, water, and ethylene glycol. An experiment on the natural convection only was conducted for cylinders with diameters between 0.03 and 0.20 mm and with aspect ratios ranging from 737 to 4790. For $5 \times 10^{-3} \leq Gr_D Pr \leq 3$, the following correlation holds

$$Nu_D = 1.02(Gr_D Pr)^{0.10} \tag{2.50}$$

Li and Parker [45], in 1967, investigated the effects of acoustics on the natural convection from horizontal wires in water. A wire with a diameter of 0.20 mm was the test specimen. For the case with no acoustic effects and for $5 \leq Gr_D Pr \leq 61$, Morgan [51] correlated the results into the following equation

$$Nu_D = 0.35(Gr_D Pr)^{0.32} \tag{2.51}$$

In 1968, Bansal and Chandna [5] published an article on free convection from horizontal cylinders. The investigators claim that the new correlation which is presented is valid for any fluid. The data are good for $Gr_D Pr$ between 10^{-5} and 10^{10}

$$[\log(Nu_D)]^2 + \left[\frac{a \log(Gr_D Pr) + d}{b} \right] \log(Nu_D)$$

$$+ \frac{1}{b} \log(Gr_D Pr)[c + \log(Gr_D Pr)] + \frac{e}{b} = 0 \tag{2.52}$$

where $a = -26.9268$, $b = 80.3767$, $c = -11.3983$, $d = 94.5623$, and $e = 1.9590$.

In 1968, Weder [71] performed natural convection experiments on horizontal cylinders in sodium hydroxide. The diameter of the cylinder was 26.2 mm, and the

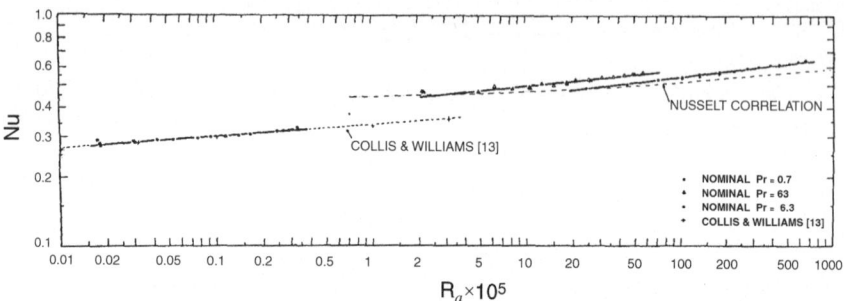

Fig. 2.1 Average Nusselt numbers versus Rayleigh number for Pr = 0.7, 6.3, and 63

aspect ratio was 4. For $6 \times 10^3 \leq \mathrm{Gr}_D \mathrm{Pr} \leq 6 \times 10^6$, the following correlation is given

$$\mathrm{Nu}_D = 0.858 (\mathrm{Gr}_D \mathrm{Pr})^{0.22} \qquad (2.53)$$

Hatton et al. [30] investigated both forced and natural convection low-speed airflow over horizontal cylinders in 1970. The cylinders tested had $0.10\ \mathrm{mm} \leq D \leq 1.26\ \mathrm{mm}$ and $96 \leq L/D \leq 1190$. The investigators, experimentally determined the average Nusselt numbers for the case of no forced airflow (natural convection only) for $4 \times 10^{-3} \leq \mathrm{Gr}_D \mathrm{Pr} \leq 10$

$$\mathrm{Nu}_D = 0.525 + 0.422 (\mathrm{Gr}_D \mathrm{Pr})^{0.315} \qquad (2.54)$$

In 1970, Gebhart, Audunson, and Pera [28, 29] published a series of articles on natural convection from long horizontal cylinders in air with a diameter of 0.01 mm and for $996 \leq L/D \leq 16200$ and for Pr = 0.7, 6.3 and 63. The experimental results are summarized in Fig. 2.1 from [29].

2.2.2 Modern Developments

Around the same time as the review article on Morgan [51] appeared, Churchill and Chu [9] published an article on laminar and turbulent natural convection from a horizontal cylinder. For the laminar regime, they took the limiting Nusselt number expressions of Saville and Churchill [63], seen in Eqs. (2.48 and 2.49), and used a form suggested by Churchill and Usagi [10] as well as considered the limiting value of the Nusselt number found by Tsubouchi and Masuda [67] in order to obtain the following expression for the average Nusselt number for horizontal cylinders

$$\mathrm{Nu}_D = 0.36 + 0.518 \left(\frac{\mathrm{Gr}_D\mathrm{Pr}}{[1 + (0.559/\mathrm{Pr})^{9/16}]^{16/9}} \right)^{1/4} \tag{2.55}$$

The preceding correlating equation matches well with experimental data for all Prandtl numbers in the range of $10^{-6} \le \mathrm{Gr}_D\mathrm{Pr} \le 10^9$. The equation, however, did not match well with the experimental data for very low values of $\mathrm{Gr}_D\mathrm{Pr} \le 10^{-6}$ which were the experimental data of Collis and Williams [13]. Another correlation, good for $10^{-11} \le \mathrm{Gr}_D\mathrm{Pr} \le 10^9$, was also developed by Churchill and Chu and is shown below

$$\mathrm{Nu}_D^{1/2} = 0.60 + 0.387 \left(\frac{\mathrm{Gr}_D\mathrm{Pr}}{[1 + (0.559/\mathrm{Pr})^{9/16}]^{16/9}} \right)^{1/6} \tag{2.56}$$

Nakai and Okazaki [52], in 1975, performed an analytical study on natural convection from small horizontal wires for low values of the Grashof number using an asymptotic matching technique. The results were compared with the experimental results of Collis and Williams [13] and valid for $10^{-9} \le \mathrm{Gr}_D \le 10^{-1}$

$$\frac{2}{\mathrm{Nu}_D} = \frac{1}{3}\ln E - \frac{1}{3}\ln\left(\frac{\mathrm{Nu}_D\mathrm{Gr}_D}{16} \right) \tag{2.57}$$

where

$$E = 3.1(\mathrm{Pr} + 9.4)^{1/2}\mathrm{Pr}^{-2} \tag{2.58}$$

In 1976, Kuehn and Goldstein [38] presented a correlation for natural convection heat transfer from a horizontal cylinder which is valid at any Rayleigh and Prandtl number

$$\frac{2}{\mathrm{Nu}_D} = \ln\left[1 + \frac{2}{\left[\left\{ 0.518\mathrm{Ra}_D^{1/4}\left[1 + \left(\frac{0.559}{\mathrm{Pr}}\right)^{3/5} \right]^{-5/12} \right\}^{15} + (0.1\mathrm{Ra}_D^{1/3})^{15} \right]^{1/15}} \right] \tag{2.59}$$

In 1977, Fand et al. [22] conducted an experimental study for natural convection heat transfer from horizontal cylinders to air, water, and silicone oils ($0.7 \le \mathrm{Pr} \le 3090$) for $2.5 \times 10^2 \le \mathrm{Gr}_D\mathrm{Pr} \le 2 \times 10^7$. The investigators correlated their data according to which temperature the fluid properties were evaluated. For the fluid properties evaluated at the mean film reference temperature, the correlation is

$$\mathrm{Nu}_D = 0.474\,\mathrm{Ra}_D^{0.25}\mathrm{Pr}^{0.047} \tag{2.60}$$

In 1979, Fujii et al. [26] performed a numerical analysis of natural convection about an isothermal horizontal cylinder for $10^{-4} \leq \mathrm{Gr}_D \leq 10^4$ and $\mathrm{Pr} = 0.7$, 10, and 100.

$$\frac{2}{\mathrm{Nu}_D} = \ln\left[1 + \frac{4.065}{C(\mathrm{Pr})\mathrm{Ra}_D^m}\right] \tag{2.61}$$

where

$$m = \frac{1}{4} + \frac{1}{10 + 4\,\mathrm{Ra}_D^{1/8}} \tag{2.62}$$

and

$$C(\mathrm{Pr}) = \frac{0.671}{[1 + (0.492/\mathrm{Pr})^{9/16}]^{4/9}} \tag{2.63}$$

The authors in [26] claim the above expression agrees well with experimental data in the range $10^{-10} \leq \mathrm{Gr}_D\mathrm{Pr} \leq 10^7$.

In 1980, Kuehn and Goldstein [39] solved the complete Navier-Stokes and energy equations for natural convection heat transfer from a horizontal isothermal cylinder, which allowed for full plume development. The authors state that previous solutions including boundary-layer assumptions and asymptotic matching solutions are not accurate over the range of $10^0 \leq \mathrm{Gr}_D\mathrm{Pr} \leq 10^7$. Kuehn and Goldstein employed a finite-difference overrelaxation technique to solve the equations numerically. The data are presented in Table 2.1. In the table, theta is the angular coordinate. See Fig. 2.2 for a schematic diagram of the relevant angles for horizontal cylinders. The data were experimentally verified by Kuehn and Goldstein.

Farouk and Güçeri [24], in 1981, numerically solved for laminar heat transfer from horizontal cylinders. They compared their results with the results of Kuehn and Goldstein [39] and found good agreement.

In 1982, Fujii et al. [27] performed analytical and experimental studies on a thin, horizontal wire. The experimental study was done on a platinum wire that was 0.470 mm in diameter and 238 and 334 mm long situated in air ($\mathrm{Pr} = 0.7$). The experimental results, for $\mathrm{Ra}_D = 0.37$, which were read from the figure in [27] are listed in Table 2.2.

Furthermore, Fujii et al. adjusted the correlation (Eqs. 2.61–2.63) found in [26] to match their present experimental data. The new correlation is as follows

$$\frac{2}{\mathrm{Nu}_D} = \ln\left[1 + \frac{3.3}{C(\mathrm{Pr})\mathrm{Ra}_D^m}\right] \tag{2.64}$$

where

$$m = \frac{1}{4} + \frac{1}{10 + 5\,\mathrm{Ra}_D^{0.175}} \tag{2.65}$$

Table 2.1 Local and average Nusselt numbers from Kuehn and Goldstein [39]

Ra_D^*	Pr	Nu_θ							Nu_D
		$\theta = 0°$	30°	60°	90°	120°	150°	180°	
10^0	0.7	1.41	1.37	1.25	1.08	0.87	0.68	0.56	1.04
10^1	0.7	1.83	1.79	1.67	1.47	1.21	0.94	0.81	1.40
10^2	0.7	2.71	2.66	2.51	2.23	1.80	1.27	0.97	2.05
10^3	0.7	3.89	3.85	3.72	3.45	2.93	2.01	1.22	3.09
10^4	0.7	6.24	6.19	6.01	5.64	4.85	3.14	1.46	4.94
10^5	0.7	10.15	10.03	9.65	9.02	7.91	5.29	1.72	8.00
10^6	0.7	16.99	16.78	16.18	15.19	13.60	9.38	2.12	13.52
10^7	0.7	29.41	29.02	27.95	26.20	23.46	16.48	2.51	23.32
10^4	0.01	3.63	3.56	3.17	2.51	1.74	1.13	0.93	2.40
10^4	0.1	5.25	5 16	4.89	4.34	3.26	1.84	1.12	3.78
10^4	1.0	6.40	6.33	6.10	5.69	4.91	3.36	1.48	5.06
10^4	5.0	6.89	6.82	6.59	6.19	5.55	4.35	1.74	5.66
10^4	10.0	7.01	6.93	6.69	6.29	5.71	4.67	1.79	5.81

Fig. 2.2 Angles of horizontal cylinders

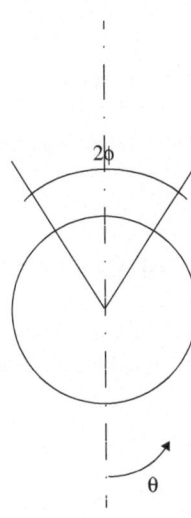

and

$$C(Pr) = \frac{0.671}{[1 + (0.492/Pr)^{9/16}]^{4/9}} \qquad (2.66)$$

In 1983, de Socio [17] experimentally investigated laminar free convection about horizontal cylinders that were partly isothermal and partly adiabatic. The motivation for this study was to determine the heat transfer around metal tubes which were

Table 2.2 Local Nusselt numbers from Fujii et al. [27]

Ra_D	Pr	Nu_θ						
		$\theta = 0°$	30°	60°	90°	120°	150°	180°
0.37	0.7	1.0	0.98	0.95	0.90	0.80	0.70	0.65

partially covered by snow or ice or around tubes that had an internal layer of deposits that affected heat transfer performance. Three cylinders were tested, each having a diameter of 37 mm and a length of 0.5 mm. Referring to Fig. 2.2, in two of the cylinders, a wedge of either $\phi = 45°$ or $90°$ was removed and replaced with an insulated Teflon (adiabatic) section. The experiments were performed for $1.5 \times 10^4 \leq Gr_D Pr \leq 6 \times 10^5$ and Pr = 0.7. The results are correlated with the following equation.

$$Nu_D = B(Gr_D Pr)^m \tag{2.67}$$

where B and m are defined in Table 2.3.

Al-Arabi and Khamis [2] experimentally determined average Nusselt numbers for air (Pr = 0.7). The isothermal boundary condition was employed using steam condensation.

For laminar flow and $1.08 \times 10^4 \leq Gr_D \leq 6.9 \times 10^5$ and $Gr_L Pr \geq 9.88 \times 10^7$, the average Nusselt number is

$$Nu_L = 0.58(Gr_L Pr)^{1/3} Gr_D^{-1/12} \tag{2.68}$$

Al-Arabi and Khamis also experimentally determined local Nusselt numbers for air.

For laminar flow and $1.08 \times 10^4 \leq Gr_D \leq 6.9 \times 10^5$ and $Gr_x Pr \geq 1.63 \times 10^8$, the local Nusselt number is

$$Nu_x = 0.58(Gr_x Pr)^{1/3} Gr_D^{-1/12} \tag{2.69}$$

Also in 1982, Farouk and Güçeri used the k-ϵ turbulence model to solve the turbulent natural convection about a horizontal cylinder. Numerical results are presented for Pr = 0.721 and $5 \times 10^7 \leq Gr_D \leq 10^{10}$. The authors compared their data to the

Table 2.3 Coefficients and exponents for Eq. (2.67)

ϕ	B	m
0 (isothermal)	0.488	0.146
45	0.543	0.239
90 (adiabatic top)	0.581	0.241
90 (adiabatic bottom)	0.569	0.236

correlations of Churchill and Chu (Eq. 2.56) and Kuehn and Goldstein (Eq. 2.59) and found good agreement with the correlation of Kuehn and Goldstein.

Wang et al. [70], in 1990, numerically computed the natural convection heat transfer from a horizontal cylinder with differing boundary condition using a spline fractional step method. Unlike previous researchers, Wang et al. solved for both the boundary layer and the resultant plume. The investigators present three cases: uniform heat flux (which will be presented in the next section), isothermal surface (which will be presented here), and mixed boundary conditions. The results for the isothermal surface are presented in Table 2.4.

The results in Table 2.4 are in good agreement with those of Kuehn and Goldstein [39] which are displayed in Table 2.1.

In 1993, Saitoh et al. [62] set forth to find benchmark solutions for the natural convection heat transfer around a horizontal isothermal cylinder. The authors of [62] employed five different kinds of numerical methodologies: (a) the ordinary explicit finite-difference method (FDM), (b) the multi-point FDM with uniform mesh, (c) the multi-point FDM with two computation domains, (d) the multi-point FDM with logarithmic coordinate transformation, and (e) the multi-point FDM with logarithmic coordinate transformation and a solid boundary condition. The results for an isothermal boundary condition are listed in Table 2.5 for the multi-point FDM with logarithmic coordinate transformation and a solid boundary condition finite-difference scheme.

The results in Table 2.5 agree well with the results of Kuehn and Goldstein and Wang et al.

Table 2.4 Local and average Nusselt numbers from Wang et al. [70]

Ra_D	Nu_θ							Nu_D
	$\theta = 0°$	30°	60°	90°	120°	150°	180°	
10^3	3.86	3.82	3.70	3.45	2.93	1.98	1.20	3.06
10^4	6.03	5.98	5.80	5.56	4.87	3.32	1.50	4.86
10^5	9.80	9.69	9.48	8.90	8.00	5.80	1.94	7.97
10^6	16.48	16.29	15.95	14.85	13.35	10.58	2.52	13.46
10^7	28.27	27.98	26.95	25.40	23.00	19.68	4.20	23.29
2×10^7	33.46	33.07	31.92	30.07	27.18	23.38	5.42	27.58

Table 2.5 Local and average Nusselt numbers from Saitoh et al. [62]

Ra_D	Nu_θ							Nu_D
	$\theta = 0°$	30°	60°	90°	120°	150°	180°	
10^3	3.813	3.772	3.640	3.374	2.866	1.975	1.218	3.024
10^4	5.995	5.935	5.750	5.410	4.764	3.308	1.534	4.826
10^5	9.675	9.557	9.278	8.765	7.946	5.891	1.987	7.898

Table 2.6 Local and average Nusselt numbers from Chouikh et al. [8]

Ra_D	Nu_θ			Nu_D
	$\theta = 0°$	$90°$	$180°$	
10^1	1.789	1.417	0.803	3.024
10^2	2.681	2.197	0.913	4.831
10^3	3.821	3.392	1.219	3.029
10^4	6.023	5.433	1.539	4.831
10^5	9.694	8.798	1.991	7.911
10^6	16.12	14.948	2.144	13.216

Chouikh et al. [8], in 1998, expanded the work of Saitoh et al. [62]. They numerically solved the conservation of mass, momentum, and energy equations via the vorticity-stream function approach. Their results for $10^1 \leq Gr_D Pr \leq 10^6$ are shown in Table 2.6.

Recently, in 2009, Atayilmaz and Teke [3] performed experimental and numerical studies on natural convection from horizontal cylinders. The authors in [3] claim that although the subject has been studied extensively for over 50 years, discrepancies in all of the data still exist due to various factors. Further, the motivating application of Atayilmaz and Teke is heat exchangers used in small refrigeration applications ($D = 4.8$ and 9.45 mm) and the authors state that there has been no investigation of this particular diameter. Based on the experimental data, the following correlation was proposed for $7.4 \times 10^1 \leq Gr_D Pr \leq 3.4 \times 10^3$ and $Pr = 0.7$

$$Nu_D = 0.954(Gr_D Pr)^{0.168} \tag{2.70}$$

Atayilmaz and Teke compared their experimental results with that of Morgan [51], Churchill and Chu [9], and Fand and Brucker [23] and found agreement to within 20 %. For the numerical portion, the authors compared their results to that of Merkin [50] who also did a numerical simulation. The two sets of results follow the same trend, but are not in good agreement.

2.3 Heat Flux Boundary Conditions

According to Dyer [16], up until his current investigation in 1965, studying natural convection from horizontal cylinders with a uniform heat flux boundary condition appeared to be neglected in the literature. Therefore, Dyer did both a theoretical and an experimental study. The theoretical study, which is valid for $10^3 \leq Gr_D^* Pr \leq 10^{10}$ yielded

$$Nu_D = 0.61(Gr_D^* Pr)^{0.192} \tag{2.71}$$

The experimental results, which were conducted in air using an electrically heated horizontal cylinder that was 7.7 cm in diameter, validated the analytical results.

In 1972, Wilks [72] did a theoretical study of natural convection from two-dimensional bodies with constant heat flux using the boundary-layer approximation. The results in [72] were correlated by Churchill [11] into the following equation

$$\mathrm{Nu}_D = 0.579 \left(\frac{\mathrm{Gr}_D \mathrm{Pr}}{[1 + (0.442/\mathrm{Pr})^{9/16}]^{16/9}} \right)^{1/4} \tag{2.72}$$

The values of the average Nusselt number were obtained by averaging the local values calculated by Wilks [72]. Churchill and Chu [12] decided to leave this expression in terms of $\mathrm{Gr}_D \mathrm{Pr}$ instead of $\mathrm{Gr}_D^* \mathrm{Pr}$ in order to show that the dependence of Nu_D on $\mathrm{Gr}_D \mathrm{Pr}$ was the same for both isothermal and uniform heat flux situations.

In 1987, Qureshi and Ahmad [58] performed a numerical solution of the full Navier-Stokes and energy equations for a uniform heat flux cylinder in air ($\mathrm{Pr} = 0.7$). The results are presented in Table 2.7 for $10^{-2} \leq \mathrm{Gr}_D^* \mathrm{Pr} \leq 10^7$.

Qureshi and Ahmad correlated the results for $10^0 \leq \mathrm{Gr}_D^* \mathrm{Pr} \leq 10^7$ and $\mathrm{Pr} = 0.7$ with the following equation.

$$\mathrm{Nu}_D = 0.800 (\mathrm{Gr}_D^* \mathrm{Pr})^{0.175} \tag{2.73}$$

For $\mathrm{Gr}_D^* \mathrm{Pr} \leq 10^0$, the average Nusselt numbers may be predicted by correlations for isothermal cylinders.

In 1990, Wang et al. [70] numerically computed the natural convection heat transfer from a horizontal cylinder with differing boundary condition using a spline fractional step method. The results for the uniform heat flux case are shown in Table 2.8.

Wang et al. [70] compared their results with Wilks [72], Churchill [11], and Qureshi and Ahmad [58] and found good agreement.

Table 2.7 Local and average Nusselt numbers from Qureshi and Ahmad [58]

Ra_D^*	Nu_θ							Nu_D
	$\theta = 0°$	30°	60°	90°	120°	150°	180°	
10^{-2}	0.48	0.47	0.47	0.46	0.53	0.45	0.45	0.46
10^{-1}	0.70	0.70	0.69	0.68	0.68	0.67	0.67	0.68
10^0	0.93	0.92	0.90	0.86	0.82	0.79	0.78	0.86
10^1	1.34	1.32	1.28	1.20	1.11	1.03	1.00	1.19
10^2	2.06	2.05	1.99	1.88	1.70	1.48	1.38	1.80
10^3	3.03	3.02	2.96	2.83	2.59	2.17	1.88	2.67
10^4	4.50	4.47	4.39	4.23	3.94	3.30	2.57	3.98
10^5	6.76	6.72	6.58	6.35	5.95	5.10	3.50	5.97
10^6	10.23	10.16	9.95	9.59	9.04	7.84	4.85	9.02
10^7	15.50	15.39	15.08	14.57	13.77	12.07	6.88	13.70

Table 2.8 Local and average Nusselt numbers from Wang et al. [70]

Ra_D^*	Nu_θ							Nu_D
	$\theta = 0°$	30°	60°	90°	120°	150°	180°	
10^6	9.87	9.83	9.60	9.24	8.94	7.91	5.02	8.88
10^7	15.04	15.00	14.72	14.08	13.58	12.28	7.14	13.57
10^8	23.12	22.84	22.57	21.92	20.85	19.63	10.87	21.00
2.5×10^8	28.05	27.34	26.98	26.18	25.01	23.17	12.26	25.08

References

1. Ackermann G (1932) Die Wärmeabgabe eines horizontalen geheizten Rohres an kaltes Wasser bei natülicher Konvecktion. Forsch Gebiete Ingenieurw 3:42–50
2. Al-Arabi M, Khamis M (1982) Natural convection heat transfer from inclined cylinders. Int J Heat Mass Transf 25:3–15
3. Atayilmaz ŞÖ, Teke I (2009) Experimental and numerical study of the natural convection from a heated horizontal cylinder. Int Commun Heat Mass 36:731–738
4. Ayrton W, Kilgour H (1892) The thermal emissivity of thin wires in air. Philos TR Soc A 183:371–405
5. Bansal T, Chandna R (1968) Empirical relationship for computation of free convection from a single horizontal cylinder to fluids. Indian J Technol 6:223–225
6. Beckers H, Haar LT, Tjoan LT, Merk H, Prins J, Schenk J (1956–1957) Heat transfer at very low Grashof and Reynolds numbers. Appl Sci Res 6:82–84
7. Champagne FH, Sleicher C, Wehrmann O (1967) Turbulence measurements with inclined hot wires. J Fluid Mech 28:153–175
8. Chouikh R, Guizani A, Maalej M, Belghith A (1998) Numerical study of the laminar natural convection flow around horizontal isothermal cylinder. Renew Energy 13:77–88
9. Churchill SW, Chu HHS (1975) Correlating equations for laminar and turbulent free convection from a vertical plate. Int J Heat Mass Transf 18:1323–1329
10. Churchill S, Usagi R (1972) A general expression for the correlation of rates of transfer and other phenomena. AICHE J 18:1121–1128
11. Churchill S (1974) Laminar free convection from a horizontal cylinder with a uniform heat flux density. Lett Heat Mass Transf 1:109–112
12. Churchill S, Chu H (1975) Correlating equations for laminar and turbulent free convection from a horizontal cylinder. Int J Heat Mass Transf 18:1049–1053
13. Collis DC, Williams M (1954) Free convection of heat from fine wires. Aeronaut Res Lab 140:1–23
14. Davis A (1922) Natural convective cooling in fluids. Phil Mag 44:920–940
15. Deaver F, Penney W, Jefferson T (1962) Heat transfer from an oscillating horizontal wire to water. J Heat Transf 84:251–256
16. Dyer J (1965) Laminar natural convection from a horizontal cylinder with a uniform convective heat flux. Trans Inst Eng Aust MC1:125–128
17. de Socio L (1983) Laminar free convection around horizontal circular cylinders. Int J Heat Mass Transf 26:1669–1677
18. Elenbaas W (1948) Dissipation of heat by free convection. Philips Res Rep 3:338–360
19. Etemad G (1955) Free convection heat transfer from a rotating horizontal cylinder to ambient air with interferometric study of flow. Trans ASME 77:1283–1289
20. Fand RM, Kaye J (1961) The influence of sound on free convection from a horizontal cylinder. J Heat Transf 83:133–148

21. Fand R, Kaye J (1963) The influence of vertical vibrations on heat transfer by free convection from a horizontal cylinder. Int Dev Heat Transfer Proc Heat Transf Conf 1961:490–498

22. Fand R, Morris E, Lum M (1977) Natural convection heat transfer from horizontal cylinders to air, water, and silicone oils for Rayleigh numbers between 3×10^2 and 2×10^7. Int J Heat Mass Transf 20:1173–1184

23. Fand R, Brucker J (1983) A correlation for heat transfer by natural convection from horizontal cylinders that accounts for viscous dissipation. Int J Heat Mass Transf 26:709–726

24. Farouk B, Güçeri SI (1981) Natural convection from a horizontal cylinder-laminar regime. J Heat Transf 103:522–527

25. Fischer J, Dosch F (1956) Die Wärmeübertagung dünner stromdurchflossener Drähte un Luft von Normalzustand bei freier Konvecktion. Z Angew Phys 8:292–297

26. Fujii T, Fujii M, Matsunaga T (1979) A numerical analysis of laminar free convection around an isothermal horizontal circular cylinder. Numer Heat Transf 2:329–344

27. Fujii T, Fujii M, Honda T (1982) Theoretical and experimental studies of the free convection around a long horizontal thin wire in air. In: Proceedings of the 7th international heat transfer conference, vol 6. Munich, Germany, pp 311–316

28. Gebhart B, Pera L (1970) Mixed convection from long horizontal cylinders. J Fluid Mech 45:49–64

29. Gebhart B, Audunson T, Pera L (1970) Forced, mixed and natural convection from long horizontal wires, experiments at various Prandtl numbers. In: Papers presented at the 4th international heat transfer conference, paper NC 3.2, Paris, France

30. Hatton A, James D, Swire H (1970) Combined forced and natural convection with low-speed air flow over horizontal cylinders. J Fluid Mech 42:17–31

31. Hermann R (1936) Wärmeübergang bei freier Strömung am waagrechten Zylinder in zweiatomigen Gasen. Ver Deut Ing Forschungsh 379

32. Jakob M, Linke W (1935) Der Wärmeübergang beim Verdampfen von Flüssigkeiten an senkrechten und waagerechten Flächen. Phys Z 36:267–280

33. Jodlbauer K (1933) Das Temperatur- und Geschwindigkeitsfeld um ein geheiztes Rohr bei freier Konvektion. Fors 4:157–172

34. Kays W, Bjorklund I (1958) Heat transfer from a rotating cylinder with and without crossflow. Trans ASME 80:70–78

35. Kennelly A, Wright C, Byvelelt JV (1909) The convection of heat from small copper wires. Trans Amer Inst Elec Eng 28:363–393

36. King W (1932) The basic laws and data of heat transmission. Mech Eng 54:347–353, 426:410–414

37. Koch W (1927) Über die Wärmeabgabe geheizter Rohre bei verschiedener Neigung der Rohrachse. Gesundh Ing Beih 1:1–29

38. Kuehn T, Goldstein R (1976) Correlating equations for natural convection heat transfer between horizontal circular cylinders. Int J Heat Mass Transf 19:1127–1134

39. Kuehn T, Goldstein R (1980) Numerical solution to the Navier-Stokes equations for laminar natural convection about a horizontal isothermal circular cylinder. Int J Heat Mass Transf 23:971–979

40. Kyte JR, Madden AJ, Piret EL (1953) Natural-convection heat transfer at reduced pressure. Chem Eng Prog 49:653–662

41. Lander C (1942) A review of recent progress in heat transfer. J Inst Mech Eng 148:81–112

42. Langmuir I (1912) The convection and conduction of heat in gases. Trans Am Inst Electr Eng 31:1229–1240

43. Lemlich R (1955) Effect of vibration on natural convective heat transfer. Ind Eng Chem 47:1175–1180

44. Lemlich R, Rao M (1965) The effect of transverse vibration on free convection from a horizontal cylinder. Int J Heat Mass Transf 8:27–33

45. Li KW, Parker JD (1967) Acoustical effects on free convective heat transfer from a horizontal wire. J Heat Transf 89:277–278

46. Mabuchi I, Tanaka T (1967–1968) Experimental study on effect of vibration on natural convective heat transfer from a horizontal fine wire. Sci Rep Res Inst Tohaku Univ Ser B 19:205–219
47. Martinelli R, Boelter L (1938) The effect of vibration on heat transfer by free convection from a horizontal cylinder. In: Proceedings of the international congress of applied mechanics
48. Mason W, Boelter L (1940) Vibration-it's effect on heat transfer. Power Plant Eng 44:43–44
49. McAdams W (1954) Heat transmission. Mc-Graw-Hill, New York
50. Merkin J (1977) Free convection boundary layers on cylinders of elliptic cross section. ASME J Heat Transf 99:453–457
51. Morgan V (1975) The overall convective heat transfer from smooth circular cylinders. Adv Heat Transf 11:199–264
52. Nakai S, Okazaki T (1975) Heat transfer form a horizontal circular wire at small Reynolds and Grashof numbers-I. Int J Heat Mass Transf 18:387–396
53. Nelson R (1924) Free convection heat in liquids. Phys Rev 23:94–103
54. Nusselt W (1929) Die Wärmeübgabe eines waagrecht liegenden Drahtes oder Rohres in Flüssigkeiten und gasen. Ver Deut Ing 73:1475–1478
55. Penney WR, Jefferson TB (1966) Heat transfer from an oscillating horizontal wire to water and ethylene glycol. J Heat Transf 88:359–366
56. Petavel J (1898) On the heat dissipated by a platinum surface at high temperatures. Philos T R Soc A 191:501–524
57. Petavel J (1901) On the heat dissipated by a platinum surface at high temperatures. Part iv. Thermal emissivity in high-pressure gases. Philos T R Soc A 197:229–254
58. Qureshi Z, Ahmad R (1987) Natural convection from a uniform heat flux horizontal cylinder at moderate Rayleigh numbers. Numer Heat Transf 11:199–212
59. Rebrov A (1961) Heat transfer with free motion about a horizontal cylinder in air. Inzh-Fiz Zh 4:32–39
60. Rice C (1923) Free and forced convection of heat in gases and liquids. Trans Am Inst Electr Eng 42:653–701
61. Rice C (1924) Free convection of heat in gases and liquids-ii. Trans Am Inst Electr Eng 43:131–144
62. Saitoh T, Sajiki T, Maruhara K (1993) Bench mark solutions to natural convection heat transfer problem around a horizontal circular cylinder. Int J Heat Mass Transf 36:1251–1259
63. Saville D, Churchill S (1967) Laminar free convection in boundary layers near horizontal cylinders and vertical axisymmetric bodies. J Fluid Mech 29:391–399
64. Schurig O, Frick C (1930) Heat and current-carrying capacity of bare conductors for outdoor service. Gen Electr Rev 33:141–157
65. Senftleben H (1951) Die Wärmeabgabe von Körpern verschiedener Form in Flüssigkeiten und Gasen bei freier Strömung. Z Angew Phys 3:361–373
66. Tsubouchi T, Sato S (1960) Heat transfer from fine wires and particles by natural convection. Chem Eng Prog Symp Ser 56:269–284
67. Tsubouchi T, Masuda H (1966–1967) Natural convection heat transfer from a horizontal circular cylinder with small rectangular grooves. Sci Rep Res Inst Tohaku Univ Ser B 18:211–242
68. van der Hegge Zijnen B (1956–1957) Modified correlation formulae for the heat transfer by forced convection from horizontal cylinders. Appl Sci Res 6:129–140
69. Wamsler F (1911) Die Wärmeabgabe geheizter Körper an Luft. Ver Deut Ing Forschunash 98/99
70. Wang P, Kahawita R, Nguyen T (1990) Numerical computation of the natural convection flow about a horizontal cylinder using splines. Numer Heat Transf 17:191–215
71. Weder E (1968) Messung des gleichzeitigen Wärme- und stoffübergangs am horizontalen Zylinder bei freier Konvecktion. Wärme Stoffübergangs 1:10–14
72. Wilks G (1972) External natural convection about two-dimensional bodies with constant heat flux. Int J Heat Mass Transf 15:351–354
73. Zhukauskas AA, Shlancyauskas A, Yaronis E (1961) Effect of ultrasonic waves on heat transfer of bodies in liquids. Inzh-Fiz Zh 4:58–62

Chapter 3
Natural Convection Heat Transfer From Vertical Cylinders

Abstract Natural convection heat transfer from vertical cylinders is an extensively studied subject. Many references provide a criterion for approximating the heat transfer from a vertical cylinder with a vertical flat plate. This can be done if the boundary-layer thickness of the resultant buoyant flow is small compared to the diameter of the cylinder. This chapter is divided into the following sections: early experimental studies, early analytical studies, and modern studies. Natural convection Nusselt number correlations for round vertical cylinders are presented.

Keywords Natural convection · Vertical cylinder · Vertical flat-plate limit · Experimental · Computational · Analytical · Nusselt numbers

3.1 Introduction

Natural convection heat transfer from vertical cylinders is a subject that has been studied extensively, but is given little attention in many well-known heat transfer textbooks (i.e., [3, 13, 17–19]). Many of these textbooks provide a criterion for approximating the heat transfer from a vertical cylinder with a vertical flat plate. This can be done if the boundary-layer thickness δ of the resultant buoyant flow is small compared to the diameter D of the cylinder. The accepted criterion provided by most heat transfer textbooks for which the flat-plate solution can be used to approximate the average Nusselt numbers for vertical cylinders within 5 % error is

$$\frac{D}{L} \geq \frac{35}{Gr_L^{0.25}} \tag{3.1}$$

In this equation, D is the diameter of the cylinder, L is the height of the cylinder, and Gr_L is the Grashof number based on the height of the cylinder.

Sparrow and Gregg [46] derived this limit in 1956 when they solved for the heat transfer and fluid flow ($Pr = 0.72$) adjacent to an isothermal cylinder. Of course, when this criterion is not met, the flat-plate approximation is not valid and the reader is directed to results by Cebeci [5] and Minkowycz and Sparrow [34]. These articles can be difficult to access and are not the only literature on the subject of vertical

© The Author(s) 2014   23
S.K.S. Boetcher, *Natural Convection from Circular Cylinders*, SpringerBriefs
in Thermal Engineering and Applied Science, DOI: 10.1007/978-3-319-08132-8_3

cylinders. The present chapter is a review of relevant literature on natural convection vertical cylinders.

3.2 Temperature Boundary Conditions

3.2.1 Early Experimenters

In 1922, Griffiths and Davis [15] experimentally determined average Nusselt numbers from isothermal vertical cylinders. Griffiths and Davis used a cylinder with a diameter of 17.43 cm and varied the length between 4.65 and 263.5 cm making the range of aspect ratios $0.267 \leq L/D \leq 15.118$. Morgan [35] correlated the data into the following equations:

For $10^7 \leq Gr_L Pr \leq 10^9$,

$$Nu_L = 0.67(Gr_L Pr)^{1/4} \qquad (3.2)$$

and for $10^9 \leq Gr_L Pr \leq 10^{11}$,

$$Nu_L = 0.0782(Gr_L Pr)^{0.357} \qquad (3.3)$$

In 1927, Koch [24] performed natural convection experiments for isothermal vertical cylinders. Koch tested four different cylinders oriented in the vertical position. The cylinders that he tested had diameters of 0.014, 0.031, 0.068, and 0.1005 m with lengths of 2.14, 1.92, 1.38, and 1.98 m, respectively. These dimensions correspond to aspect ratios between 20 and 153. The results reported are based on the diameter of the cylinder and are inconclusive.

Jakob and Linke [21, 35] experimentally determined the average Nusselt numbers for natural convection for vertical cylinders situated in air in 1935. The cylinder used in the study had a diameter of 35 mm and an aspect ratio $L/D = 4.3$.

For laminar flow $10^4 \leq Gr_L Pr \leq 10^8$,

$$Nu_L = 0.56(Gr_L Pr)^{1/4} \qquad (3.4)$$

and for turbulent flow $10^8 \leq Gr_L Pr \leq 10^{12}$,

$$Nu_L = 0.13(Gr_L Pr)^{1/3} \qquad (3.5)$$

In 1937, Carne [4] performed experiments of heated cylinders in air for 0.475 cm $\leq D \leq 7.62$ cm and $8 \leq L/D \leq 127$. The boundary condition at the surface was assumed isothermal. Morgan [35] correlated the results. For $2 \times 10^6 \leq Gr_L Pr \leq 2 \times 10^8$,

$$Nu_L = 1.07(Gr_L Pr)^{0.28} \qquad (3.6)$$

and for $2 \times 10^8 \leq Gr_L Pr \leq 2 \times 10^{11}$,

$$Nu_L = 0.152(Gr_L Pr)^{0.38} \tag{3.7}$$

Eigenson [9], in 1940, performed natural convection experiments in air (diatomic gases) on cylinders with a height of 2.5 m and diameters of 2.4, 7.55, 15, 35, and 50 mm. He also performed an experiment on a cylinder with a diameter of 58 mm and a length of 6.5 m. Eigenson determined that when $Gr_D \geq 10^6$, curvature effects became negligible; therefore, the same average Nusselt number correlations could be applied to both cylinders and flat plates.

The following results of Eigenson are for $Gr_D \geq 10^6$, which the author claims is appropriate for both cylinders and flat plates. For $Gr_L \leq 10^9$,

$$Nu_L = 0.48(Gr_L)^{1/4} \tag{3.8}$$

and for the transitional region $10^9 \leq Gr_L \leq 1.69 \times 10^{10}$,

$$Nu_L = 51.5 + 0.0000726(Gr_L)^{0.63} \tag{3.9}$$

and for the turbulent region $Gr_L \geq 1.69 \times 10^{10}$,

$$Nu_L = 0.148(Gr_L)^{1/3} - 127.6 \tag{3.10}$$

In 1942, Mueller [36] experimentally determined natural convection Nusselt numbers for vertical wires. The temperature difference between the wire and the tube was varied parametrically. Interestingly, it was found that the length of the wire had no effect on the heat transfer coefficients; thus, Mueller based the results on the wire diameter. Morgan [35] correlated the results in [36] into the following equation for $10^{-6} \leq Gr_D Pr \leq 10^{-2}$

$$Nu_D = 1.0(Gr_D Pr)^{0.11} \tag{3.11}$$

In 1948, Touloukian et al. [47] experimentally investigated natural convection heat transfer from vertical cylindrical surfaces in a liquid environment, as opposed to a gas environment. It is interesting to note that the authors of [47] actually wanted to use a plane surface, but ended up using a cylindrical surface for several reasons. First, a cylindrical test section was easier to prepare (i.e., it can be taken from standard pipes) than a rectangular test section. Second, a cylindrical test section has edge losses on two ends only as opposed to four ends like a rectangular test section. Third, uniform heating was easier, and finally, the cylinder would require a much smaller liquid bath. Tests were conducted using 2.75-inch-diameter cylinders with heights of 6, 18, and 36.25 in. In order to achieve a uniform temperature at the surface of the cylinder, the investigators varied the heat along the length of the cylinder by employing separate, controllable heating sections.

For water and ethylene-glycol and laminar flow $2 \times 10^8 \leq Gr_L Pr \leq 4 \times 10^{10}$,

$$Nu_L = 0.726(Gr_L Pr)^{1/4} \tag{3.12}$$

and for the same fluids and turbulent flow $4 \times 10^{10} \leq Gr_L Pr \leq 9 \times 10^{11}$,

$$Nu_L = 0.0674(Gr_L Pr^{1.29})^{1/3} \tag{3.13}$$

In 1953, Kyte et al. [26] experimentally investigated the heat transfer from spheres and horizontal cylinders at low pressures. In addition, they were interested in seeing the effect of orientation of the cylinder on heat transfer, so they also ran studies on a 0.00306-in. diameter, 16.62-in.-long vertical cylinder in air. For $10^{-11} \leq Gr_D Pr D/L \leq 10^{-4.5}$,

$$Nu_D = \frac{2}{\ln\left[1 + \frac{4.47}{(Gr_D Pr \frac{D}{L})^{0.26}}\right]} \tag{3.14}$$

In 1954, McAdams [32] correlated the experimental data of Saunders [44] and Wiese [48]. The experimental data of Saunders and Wiese were originally performed for a heated flat plate in air, but according to McAdams, it can be also used for a vertical cylinder. For the laminar range $10^4 \leq Gr_L Pr \leq 10^9$, the data can be represented by

$$Nu_L = 0.59(Gr_L Pr)^{1/4} \tag{3.15}$$

For $Gr_L Pr \leq 10^4$, a correlation equation is not given. Instead, the curve fit is displayed on a graph and the coordinates of the graph are given in Table 3.1 and for turbulent flow, $10^9 \leq Gr_L Pr \leq 10^{12}$

$$Nu_L = 0.13(Gr_L Pr)^{1/3} \tag{3.16}$$

In 1958, Kreith [25] correlated experimental heat transfer data for isothermal vertical cylinders. For laminar flow, $10^5 \leq Gr_L Pr \leq 10^9$, the average Nusselt number is reported as

$$Nu_L = 0.555(Gr_L Pr)^{1/4} \tag{3.17}$$

Table 3.1 Curve coordinates for $Gr_L Pr \leq 10^4$ [32]

$Gr_L Pr$	Nu_L
10^0	1.44
10^1	1.90
10^2	2.63
10^3	3.89
10^4	6.03

and for turbulent flow, $10^9 \leq Gr_L Pr \leq 10^{12}$

$$Nu_L = 0.0210(Gr_L Pr)^{2/5} \tag{3.18}$$

In 1969, Nagendra et al. [37] experimentally determined average Nusselt numbers for isothermal vertical cylinders in water $(Pr = 5)$. They performed two sets of experiments: one with thin wires of length 250 mm and diameters 0.5, 1.0, and 1.25 mm, and one larger cylinder with a diameter of 0.315 in. and length 12 in. The authors state that while the condition of constant heat flux for thin wires translates to isothermal walls, the condition of constant heat flux on the 0.315-inch-diameter cylinder leads to a 20 % variation in temperature along the wall. The experimental results obtained in this study agree with the analytical Nusselt number correlations by the same authors, which appear in a later paper [38] and will be discussed later.

In 1970, Hanesian and Kalish [16] performed natural convection experiments on heated cylinders in air and fluorocarbon gases (CF_2Cl_2, CF_3Cl, CF_4, CHF_2Cl, CHF_3, C_2F_6, C_3F_8, and c-C_4F_8). The authors used a cartridge heater and assumed a uniform surface temperature. The test cylinder had a diameter of 1 in. and a length of 3.125 in. The empirical results were correlated by Morgan [35]. For $10^6 \leq Gr_L Pr \leq 10^8$,

$$Nu_L = 0.48(Gr_L Pr)^{0.23} \tag{3.19}$$

In 1970, Fujii et al. [12] experimentally investigated the natural convection heat transfer from the outer surface of a vertical cylinder in water, spindle oil, and mobiltherm oil ($2 \leq Pr \leq 200$). The diameter of the cylinder was 82 mm and the height of the cylinder was 1 m; therefore, $L/D = 12.2$. Fujii et al. also witnessed four regions of boundary-layer development: laminar, vortex street, transition turbulent, and fully turbulent. The investigators found local Nusselt numbers for both uniform wall temperature and uniform wall heat flux. They also found average Nusselt numbers for uniform wall temperature. Fujii et al. found that the experimental results for the curved surfaces were approximately 1.3 % larger than the equivalent results for a flat plate.

The investigators in [12] correlated their data using the following reference temperatures

$$T_m = \frac{1}{2}(T_{\text{cylinder}} + T_\infty) \tag{3.20}$$

and

$$T_e = T_{\text{cylinder}} - \frac{1}{4}(T_{\text{cylinder}} + T_\infty) \tag{3.21}$$

where T_m is the average temperature, T_{cylinder} is the surface temperature, T_∞ is the ambient temperature, and T_e is a reference temperature.

Fujii et al. formed the following correlations based on the dimensionless numbers being evaluated at the reference temperature T_e. This is indicated by the notation $(\ )_e$. The only exception to this is that the coefficient of thermal expansion β is to be evaluated on the basis of T_m. The following represents local Nusselt numbers for the isothermal wall temperatures case. The data for the uniform wall heat flux case are presented in the next section.

Water (Pr = 5)
For laminar flow $10^8 \leq (Gr_x Pr)_e \leq (1\text{--}3) \times 10^{10}$,

$$(Nu_x)_e = 0.45(Gr_x Pr)_e^{1/4} \tag{3.22}$$

and for turbulent flow $(4\text{--}8) \times 10^{10} \leq (Gr_x Pr)_e \leq 2 \times 10^{12}$,

$$(Nu_x)_e = 0.130(Gr_x Pr)_e^{1/3} \tag{3.23}$$

Spindle Oil (Pr = 100)
For laminar flow and $10^9 \leq (Gr_x Pr)_e \leq (3\text{--}7) \times 10^{10}$,

$$(Nu_x)_e = 0.49(Gr_x Pr)_e^{1/4} \tag{3.24}$$

and for turbulent flow $8 \times 10^{10} \leq (Gr_x Pr)_e \leq 6 \times 10^{11}$,

$$(Nu_x)_e = 0.87(Gr_x Pr)_e^{1/4} \tag{3.25}$$

and for $6 \times 10^{11} \leq (Gr_x Pr)_e \leq 10^{13}$,

$$(Nu_x)_e = 0.0150(Gr_x Pr)_e^{2/5} \tag{3.26}$$

Mobiltherm Oil (Pr = 100)
For laminar flow and $10^9 \leq (Gr_x Pr)_e \leq (1.5\text{--}9) \times 10^{11}$,

$$(Nu_x)_e = 0.49(Gr_x Pr)_e^{1/4} \tag{3.27}$$

and for turbulent flow $(2\text{--}4) \times 10^{11} \; (Gr_x Pr)_e \leq 3 \times 10^{12}$,

$$(Nu_x)_e = 0.87(Gr_x Pr)_e^{1/4} \tag{3.28}$$

Since kinematic viscosity is highly dependent on temperature, Fujii et al. recast Eqs. (3.22–3.28) (except Eqs. (3.25) and (3.26) for turbulent spindle oil which will be shown below) in terms of specified kinematic viscosity (ν_s is the kinematic viscosity at the surface temperature and ν_∞ is the kinematic viscosity at the ambient temperature) and neglecting variation of other material properties. The recast equations are arrived at by replacing

$$(Nu_x)_e \rightarrow (Nu_x)_\infty (\nu_s/\nu_\infty)^{0.21} \tag{3.29}$$

and

$$(Gr_x Pr)_e \rightarrow (Gr_x Pr)_\infty \tag{3.30}$$

with the exception of Eq. (3.25) which is for spindle oil and $2.0 \times 10^{11} \leq (Gr_x Pr)_\infty \leq 2.0 \times 10^{12}$

$$(Nu_x)_\infty (v_s/v_\infty)^{0.21} = 0.0175 (Gr_x Pr)_\infty^{2/5} \tag{3.31}$$

In addition to determining local heat transfer coefficients, Fujii et al. also calculated the average heat transfer coefficients.

Water (Pr = 5)
For $(1.5–4) \times 10^{10} \leq (Gr_L Pr)_\infty \leq 10^{12}$

$$(Nu_L)_\infty (v_s/v_\infty)^{0.21} = 0.130 (Gr_L Pr)_\infty^{1/3} - (111–176) \tag{3.32}$$

Spindle Oil and Mobiltherm Oil (Pr = 100)
For $(1.5–4) \times 10^{10} \leq (Gr_L Pr)_\infty \leq 2.0 \times 10^{11}$,

$$(Nu_L)_\infty (v_s/v_\infty)^{0.21} = 1.16 (Gr_L Pr)_\infty^{1/4} - (117–227) \tag{3.33}$$

For $2.0 \times 10^{11} \leq (Gr_L Pr)_\infty \leq 10^{12}$,

$$(Nu_L)_\infty (v_s/v_\infty)^{0.21} = 0.0145 (Gr_L Pr)_\infty^{2/5} + (115–65) \tag{3.34}$$

Finally, Fujii et al. suggest an equation for the average Nusselt number that can be used for water, spindle oil, or mobiltherm oil. This is the correlation that is found in Morgan [35]. For $10^{10} \leq (Gr_L Pr)_\infty \leq 10^{12}$,

$$(Nu_L)_\infty (v_s/v_\infty)^{0.21} = (0.017 \pm 0.002)(Gr_L Pr)_\infty^{2/5} \tag{3.35}$$

3.2.2 Early Analytical

The earliest analytical prediction of Nusselt numbers from isothermal vertical cylinders was by Elenbaas [10] in 1948. Elenbaas analytically determined these Nusselt numbers by assuming that the natural convection problem may be solved using heat conduction, which was proposed by Langmuir [27]. The correlation proposed by Elenbaas was

$$Nu_D \cdot \exp\left[-\frac{2}{Nu_D}\right] = 0.6 \left(\frac{D}{L} Gr_D Pr\right)^{1/4} \tag{3.36}$$

In 1951, Senftleben [45] analytically determined average Nusselt numbers for laminar isothermal vertical cylinders.

$$Nu_L = \frac{2L}{\varphi D} \left[1 - \frac{0.102}{\varphi (Gr_L Pr)^{1/4}} \left\{ \left(1 + \frac{\varphi (Gr_L Pr)^{1/4}}{0.130} \right)^{1/2} - 1 \right\} \right] \qquad (3.37)$$

where

$$\varphi = \ln \left[1 + \frac{L}{D} \frac{3.34}{(Gr_L Pr)^{1/4}} \right] \qquad (3.38)$$

As mentioned previously, Sparrow and Gregg [46] used a pseudosimilarity variable coordinate transformation and perturbation technique to solve for the heat transfer and fluid flow field adjacent to an isothermal vertical cylinder for $Pr = 0.72$ and 1 and $0 < \xi < 1$, where ξ is the curvature parameter defined as

$$\xi = \frac{4L}{D} \left(\frac{Gr_L}{4} \right)^{-1/4} \qquad (3.39)$$

Sparrow and Gregg used the boundary-layer approximation. The results in [46] were presented in graphical form.

Also in 1956, LeFevre and Ede [30] (and later published again in [8]), used an integral method to obtain the following Nusselt number correlations

$$Nu_x = \left[\frac{7 Gr_x Pr^2}{5(20 + 21 Pr)} \right]^{1/4} + \frac{4(272 + 315 Pr)x}{35(64 + 63 Pr)D} \qquad (3.40)$$

and

$$Nu_L = \frac{4}{3} \left[\frac{7 Gr_L Pr^2}{5(20 + 21 Pr)} \right]^{1/4} + \frac{4(272 + 315 Pr)L}{35(64 + 63 Pr)D} \qquad (3.41)$$

Here, Nu_x is the local Nusselt number, Nu_L is the average Nusselt number and Pr is the Prandtl number.

Millsaps and Pohlhausen [33] analytically determined average Nusselt numbers for isothermal vertical cylinders for $0.733 \le Pr \le 100$ using boundary-layer theory. Originally, Millsaps and Pohlhausen presented their results in terms of Nusselt and Grashof numbers in terms of the radius of the cylinder. Those original results have been converted into values based on the diameter of the cylinder and presented in Table 3.2.

In 1970, Fujii and Uehara [11] solved for the local Nusselt number for a vertical cylinder when the equivalent local Nusselt number for a flat plate with the same boundary condition is known. For any Pr number and when the surface temperature is given,

$$Nu_x = Nu_{x,fp} + 0.870 \frac{x}{D}, \quad \frac{x}{D} \le 0.35 \, Nu_{x,fp} \qquad (3.42)$$

Table 3.2 Average Nusselt
numbers from Millsaps and
Pohlhausen [33]

Pr	Gr_D	Nu_D
0.733	400	3.704
	800	4.254
	4,000	5.956
	8,000	6.922
	40,000	9.920
	80,000	11.636
	400,000	16.958
	800,000	20.000
1	800	4.632
	8,000	7.588
	80,000	12.814
	800,000	22.100
10	800	8.450
	8,000	14.328
	80,000	24.760
	800,000	43.300
100	800	14.932
	8,000	25.820
	80,000	45.180
	800,000	79.120

Also in 1970, Negendra et al. [38] analytically determined average Nusselt numbers for vertical cylinders with a power-law surface temperature variation of the
form

$$T_{\text{cylinder}} = T_\infty + N(x/L)^n \tag{3.43}$$

where N is a constant, n is an exponent, and x is the vertical distance measured from
the base of the cylinder. The results were correlated of the form

$$Nu_D = C_1 (Ra_D D/L)^{a_1} \tag{3.44}$$

where C_1 and a_1 are defined in Table 3.3 for three values of $n = 2.0, 0.0$, and -0.5
in Eq. (3.43). When $n = 0.0$, the wall temperature is uniform.

Experimental data obtained by the same authors in [37] corroborate the analytical
findings.

A companion article to [38] by Nagendra et al. [40] examined the effect of exponential surface temperature variation on heat transfer coefficients. The surface
temperature variation is given by

$$T_{\text{cylinder}} = T_\infty + M \exp(mx) \tag{3.45}$$

Table 3.3 Constants for Nagendra et al. [38] correlation

n (Eq. 3.43)		Short cylinders (flat plates) ($Ra_D D/L \geq 10^4$)	Long cylinders ($0.05 \leq Ra_D D/L \leq 10^4$)	Wires ($Ra_D D/L \leq 0.05$)
2.0	C_1	0.97	1.75	1.16
	a_1	0.25	0.18	0.05
0.0	C_1	0.57	1.30	0.87
	a_1	0.25	0.16	0.05
−0.5	C_1	0.25	1.30	1.23
	a_1	0.25	0.08	0.05

where M and m are constants. The isothermal case corresponds to $m = 0$. The results were determined analytically and correlated into the following equations. The $Ra_D D/L$ for short cylinders, long cylinders, and wires is the same as above.

Short Cylinders (Flat Plates)

$$Nu_D = 0.97 \left(\frac{e^m - e^{-0.25m}}{m^{0.75}} \right) (Ra_D D/L)^{0.25} \tag{3.46}$$

Long Cylinders

$$Nu_D = 1.86 \left(\frac{e^m - e^{-0.16m}}{m^{0.84}} \right) (Ra_D D/L)^{0.16} \tag{3.47}$$

Wires

$$Nu_D = 1.074 \left(\frac{e^m - e^{-0.05m}}{m^{0.95}} \right) (Ra_D D/L)^{0.05} \tag{3.48}$$

In 1974, Cebeci [5] extended the work of Sparrow and Gregg [46] to include $0.01 \leq Pr \leq 100$ and $0 < \xi < 5$. Cebeci numerically solved the boundary-layer equations using a two-point finite-difference method. The local Nusselt numbers obtained by Cebeci are shown in Table 3.4, and the average Nusselt numbers are shown in Table 3.5. In the tables, $Nu_{x,fp}$ and $Nu_{L,fp}$ are the local and average Nusselt numbers for an isothermal flat plate.

The local Nusselt number for a flat plate was calculated by Ostrach [41] and correlated by LeFevre [29]

$$Nu_{x,fp} = Gr_x^{1/4} \frac{0.75 Pr^{1/2}}{(0.609 + 1.221 Pr^{1/2} + 1.238 Pr)^{1/4}} \tag{3.49}$$

Table 3.4 Local Nusselt numbers from Cebeci [5]

ξ	$Nu_x/Nu_{x,fp}$					
	$Pr = 0.01$	$Pr = 0.10$	$Pr = 0.72$	$Pr = 1.0$	$Pr = 10$	$Pr = 100$
0	1.000	1.000	1.000	1.000	1.000	1.000
0.159	1.397	1.148	1.069	1.062	1.031	1.016
0.283	1.646	1.251	1.121	1.108	1.055	1.029
0.503	2.087	1.429	1.210	1.188	1.096	1.051
0.752	2.561	1.619	1.307	1.276	1.141	1.076
1.064	3.124	1.843	1.422	1.380	1.196	1.107
1.337	3.609	2.033	1.521	1.469	1.243	1.133
1.480	3.859	2.129	1.571	1.515	1.268	1.147
1.891	4.566	2.387	1.711	1.643	1.339	1.185
2.093	4.909	2.512	1.778	1.704	1.373	1.204
2.378	5.387	2.684	1.871	1.788	1.420	1.230
2.632	5.808	2.834	1.952	1.862	1.461	1.253
2.828	6.131	2.949	2.013	1.918	1.492	1.271
3.364	7.000	3.255	2.177	2.068	1.575	1.318
3.722	7.577	3.455	2.285	2.166	1.630	1.349
4.000	8.018	3.608	2.366	2.240	1.672	1.373
4.229	8.381	3.733	2.432	2.301	1.707	1.392
4.681	9.087	3.976	2.561	2.419	1.773	1.431
5.030	9.629	4.162	2.660	2.508	1.823	1.460

and the average Nusselt number is usually found by using the correlation of Churchill and Chu [6], which is shown below

$$Nu_{L,fp} = 0.68 + \frac{0.670 Ra_L^{1/4}}{[1 + (0.492/Pr)^{9/16}]^{4/9}} \tag{3.50}$$

Minkowycz and Sparrow [34] investigated the impact of different levels of truncation in the series solution of Sparrow and Gregg [46] in 1974. The findings in [34] are presented graphically for $0 < \xi < 10$ and for $Pr = 0.733$ and are in good agreement with [46].

3.2.3 Modern Developments

Al-Arabi and Khamis [1] experimentally determined average Nusselt numbers for air ($Pr = 0.7$) in 1982. The isothermal boundary condition was employed using steam condensation.

Table 3.5 Average Nusselt numbers from Cebeci [5]

ξ	$Nu_L/Nu_{L,fp}$					
	$Pr = 0.01$	$Pr = 0.10$	$Pr = 0.72$	$Pr = 1.0$	$Pr = 10$	$Pr = 100$
0	1.000	1.000	1.000	1.000	1.000	1.000
0.159	1.322	1.118	1.054	1.048	1.024	1.012
0.283	1.497	1.188	1.090	1.081	1.041	1.021
0.503	1.805	1.310	1.153	1.138	1.071	1.038
0.752	2.151	1.456	1.224	1.202	1.105	1.056
1.064	2.581	1.613	1.312	1.281	1.147	1.079
1.337	2.953	1.758	1.387	1.350	1.184	1.099
1.480	3.147	1.832	1.426	1.385	1.203	1.110
1.891	3.698	2.043	1.536	1.485	1.257	1.139
2.093	3.965	2.145	1.590	1.533	1.283	1.154
2.378	4.340	2.286	1.664	1.601	1.319	1.174
2.632	4.671	2.410	1.729	1.660	1.351	1.192
2.828	4.924	2.504	1.779	1.705	1.376	1.205
3.364	5.607	2.755	1.911	1.825	1.441	1.242
3.722	6.056	2.918	1.997	1.903	1.484	1.267
4.000	6.400	3.042	2.063	1.963	1.517	1.285
4.229	6.681	3.142	2.116	2.011	1.544	1.300
4.681	7.228	3.334	2.180	2.104	1.596	1.330
5.030	7.644	3.478	2.295	2.175	1.636	1.352

For laminar flow and $1.08 \times 10^4 \leq Gr_D \leq 6.9 \times 10^5$ and $9.88 \times 10^7 \leq Gr_L Pr \leq 2.6 \times 10^9$, the average Nusselt number is

$$Nu_L = 2.9(Gr_L Pr)^{1/4} Gr_D^{-1/12} \qquad (3.51)$$

and for turbulent flow when $1.08 \times 10^4 \leq Gr_D \leq 6.9 \times 10^5$ and $2.6 \times 10^9 \leq Gr_L Pr \leq 2.95 \times 10^{10}$

$$Nu_L = 0.47(Gr_L Pr)^{1/3} Gr_D^{-1/12} \qquad (3.52)$$

Al-Arabi and Khamis also experimentally determined local Nusselt numbers for air.

For laminar flow and $1.08 \times 10^4 \leq Gr_D \leq 6.9 \times 10^5$ and $1.63 \times 10^8 \leq Gr_x Pr \leq 2.16 \times 10^9$, the local Nusselt number is

$$Nu_x = 2.3(Gr_x Pr)^{1/4} Gr_D^{-1/12} \qquad (3.53)$$

and for turbulent flow when $1.08 \times 10^4 \leq Gr_D \leq 6.9 \times 10^5$ and $4.4 \times 10^9 \leq Gr_x Pr \leq 2.3 \times 10^{10}$

$$Nu_x = 0.42(Gr_x Pr)^{1/3} Gr_D^{-1/12} \tag{3.54}$$

In 1988, Lee et al. [28] used the similarity solution to solve the boundary-layer equations for nonuniform wall temperature. The work in [28] extended the work of Fujii and Uehara [11]. Local and average Nusselt numbers are calculated for $0 < \xi < 70$ and $0.1 \le Pr \le 100$. The investigators solved for wall temperature of the form

$$T_{\text{cylinder}} = T_\infty + ax^n \tag{3.55}$$

where T_{cylinder} is the temperature at the surface of the cylinder, T_∞ is the ambient temperature, a is a dimensional constant, and n is a constant exponent. When $n = 0$, the cylinder wall temperature is uniform.

The local Nusselt number is

$$\ln\left[Nu_x\left(\frac{Gr_x}{4}\right)^{-1/4}\right] = R_1(\xi) + \left\{\ln\left[Nu_{x,fp}\left(\frac{Gr_x}{4}\right)^{-1/4}\right] - R_1(0)\right.$$

$$+ 0.97031n - 1.22087n^2 + 0.64755n^3\Bigg\}$$

$$* exp(-p_1\xi^{1/2}(1 - 0.39086n + 0.510111n^2 - 0.25295n^3)) \tag{3.56}$$

and R_1 is a function such that

$$R_1(\xi) = -2.92620 + 1.66850\xi^{1/2} - 0.21909\xi + 0.011308\xi^{3/2} \tag{3.57}$$

and p_1 is defined as

$$p_1 = 0.34181 + 0.36050 Pr^{-0.24563} \tag{3.58}$$

with $Nu_{x,fp}$ is

$$Nu_{x,fp}\left(\frac{Gr_x}{4}\right)^{-1/4} = (3/4)(2Pr)^{1/2}[2.5(1 + 2Pr^{1/2} + 2Pr)]^{-1/4} \tag{3.59}$$

and the average Nusselt number is

$$\ln\left[Nu_L\left(\frac{Gr_L}{4}\right)^{-\frac{1}{4}}\right] = R_2(\xi) + \left\{\ln\left[Nu_{L,fp}\left(\frac{Gr_L}{4}\right)^{-\frac{1}{4}}\right] - R_2(0)\right.$$

$$+ 0.97034n - 1.22080n^2 + 0.64746n^3\Bigg\}$$

$$* exp(-p_2\xi^{1/2}(1 - 0.24592n + 0.45745n^2 - 0.25153n^3)) \tag{3.60}$$

and R_2 is a function such that

$$R_2(\xi) = -2.92620 + 1.66850\xi^{1/2} - 0.21909\xi + 0.011308\xi^{3/2} \qquad (3.61)$$

and

$$p_2 = 0.29369 + 0.32635Pr^{-0.19305} \qquad (3.62)$$

with $Nu_{L,fp}$ being defined as

$$Nu_{L,fp}\left(\frac{Gr_L}{4}\right)^{-1/4} = (2Pr)^{1/2}[2.5(1 + 2Pr^{1/2} + 2Pr)]^{-1/4} \qquad (3.63)$$

In 2007, Popiel et al. [43] experimentally investigated natural convection from an isothermal vertical cylinder for $Pr = 0.71$, $1.5 \times 10^8 < Ra_L < 1.1 \times 10^9$, and $0 < L/D < 60$. The top of the cylinder was insulated, and the cylinder was situated on an insulated surface. The experimental results were correlated into the following equation

$$Nu_L = ARa_L^n \qquad (3.64)$$

where

$$A = 0.519 + 0.03454\frac{L}{D} + 0.0008772\left(\frac{L}{D}\right)^2 + 8.855 \times 10^{-6}\left(\frac{L}{D}\right)^3 \qquad (3.65)$$

and

$$n = 0.25 - 0.00253\frac{L}{D} + 1.152 \times 10^{-5}\left(\frac{L}{D}\right)^2 \qquad (3.66)$$

Popiel et al. found that their results agreed with Cebeci [5].

Furthermore, Popiel [42] took the tabular results from Cebeci [5] and formulated correlating equations for both the local and average Nusselt numbers for $Pr = 0.72$

$$\frac{Nu_x}{Nu_{x,fp}} = 1 + 0.400\left[32^{0.5}Gr_x^{-0.25}\frac{x}{D}\right]^{0.886} \qquad (3.67)$$

$$\frac{Nu_L}{Nu_{L,fp}} = 1 + 0.300\left[32^{0.5}Gr_L^{-0.25}\frac{L}{D}\right]^{0.909} \qquad (3.68)$$

and for $Pr = 6$ (for water at 25 °C),

$$\frac{Nu_x}{Nu_{x,fp}} = 1 + 0.214\left[32^{0.5}Gr_x^{-0.25}\frac{x}{D}\right]^{0.914} \qquad (3.69)$$

$$\frac{Nu_L}{Nu_{L,fp}} = 1 + 0.160 \left[32^{0.5} Gr_L^{-0.25} \frac{L}{D} \right]^{0.931} \tag{3.70}$$

Popiel also provides a slightly less accurate correlation for all Prandtl numbers $0.01 \le Pr \le 100$

$$\frac{Nu_L}{Nu_{L,fp}} = 1 + B \left[32^2 Gr_L^{-0.25} \frac{L}{D} \right]^C \tag{3.71}$$

where

$$B = 0.0571322 + 0.20305 Pr^{-0.43} \tag{3.72}$$

and

$$C = 0.9165 - 0.0043 Pr^{0.5} + 0.013333 \ln Pr + 0.0004809/Pr \tag{3.73}$$

Furthermore, Popiel also used the results of Cebeci to formulate a new criterion in which curvature effects cause Nusselt numbers to deviate from the flat-plate solution. Cebeci claims that unlike Eq. (3.1), which is accurate to 5 %, the new equation, shown below, is within 3 %.

$$Gr_L^{0.25} \frac{D}{L} \ge a + \frac{b}{Pr^{0.5}} + \frac{c}{Pr^2} \tag{3.74}$$

where $a = 11.474$, $b = 48.92$, and $c = -0.006085$.

In 2013, Day et al. [7] examined the subject of natural convection from isothermal vertical cylinders. The authors in [7] performed numerical simulations for $10^2 < Ra_L < 10^9$, $0.1 < L/D < 10$, and $Pr = 0.7$ and compared those results to several other widely used correlations. The results were correlated into the following equations
For $0.1 \le L/D \le 1$,

$$Nu_L = -0.2165 + 0.5204 Ra_L^{1/4} + 0.8473 \left[\frac{L}{D} \right] \tag{3.75}$$

For $2 \le L/D \le 10$,

$$Nu_L = -0.06211 + 0.5441 Ra_L^{1/4} + 0.6123 \left[\frac{L}{D} \right] \tag{3.76}$$

3.3 Heat Flux Boundary Conditions

As was mentioned in the section on temperature boundary conditions, Fujii et al. [12] performed natural convection experiments on vertical cylinders. In addition to testing isothermal boundary conditions, they also tested uniform heat flux. As a reminder, the Nusselt numbers were correlated using the reference temperatures in Eqs. (3.20) and (3.21).

Water ($Pr = 5$)

For uniform heat flux and laminar flow $10^{10} \leq (Gr_x^* Pr)_e \leq (0.2\text{--}2) \times 10^{13}$,

$$(Nu_x)_e = 0.58(Gr_x^* Pr)_e^{1/5} \tag{3.77}$$

and turbulent flow $(1\text{--}7) \times 10^{13} \leq (Gr_x^* Pr)_e \leq 2 \times 10^{15}$

$$(Nu_x)_e = 0.215(Gr_x^* Pr)_e^{1/4} \tag{3.78}$$

Spindle Oil ($Pr = 100$)

For laminar flow and $10^{11} \leq (Gr_x^* Pr)_e \leq (0.5\text{--}3) \times 10^{13}$,

$$(Nu_x)_e = 0.62(Gr_x^* Pr)_e^{1/5} \tag{3.79}$$

for turbulent flow $(4\text{--}8) \times 10^{13} \leq (Gr_x^* Pr)_e \leq 5 \times 10^{14}$,

$$(Nu_x)_e = 0.90(Gr_x^* Pr)_e^{1/5} \tag{3.80}$$

and for $5 \times 10^{14} \leq (Gr_x^* Pr)_e \leq 2 \times 10^{16}$,

$$(Nu_x)_e = 0.050(Gr_x^* Pr)_e^{2/7} \tag{3.81}$$

Mobiltherm Oil ($Pr = 100$)

For laminar flow and $10^{10} \leq (Gr_x^* Pr)_e \leq (0.6 \sim 6) \times 10^{14}$,

$$(Nu_x)_e = 0.62(Gr_x^* Pr)_e^{1/5} \tag{3.82}$$

for turbulent flow and $(2\text{--}4) \times 10^{14} \leq (Gr_x^* Pr)_e \leq 3 \times 10^{15}$,

$$(Nu_x)_e = 0.90(Gr_x^* Pr)_e^{1/5} \tag{3.83}$$

In 1970, Fujii and Uehara [11], for any Prandtl number, analytically determined the local Nusselt number for a uniform heat flux.

$$Nu_x = Nu_{x,fp} + 0.690\frac{x}{D}, \quad \frac{x}{D} \leq 0.35 Nu_{x,fp} \tag{3.84}$$

Also in 1970, Nagendra et al. [39] analytically determined the average Nusselt numbers for vertical cylinders with uniform heat flux for three different categories: short cylinders, long cylinders, and wires. Note that the authors in [39] calculated average Nusselt numbers based on the diameter D of the cylinder and not the height

L. The Nusselt number, for any Prandtl number, based on the modified Rayleigh number Ra_D^* is of the form

$$Nu_D = C_1 \left(Ra_D^* \frac{D}{L} \right)^{C_2}$$ (3.85)

Nagendra et al. also put the equation in the form of the traditional diameter-based Rayleigh number by replacing $T_{cylinder} - T_\infty$ with $\overline{T_{cylinder} - T_\infty}$. The equation is of the form

$$Nu_D = C_3 \left(Ra_D \frac{D}{L} \right)^{C_4}$$ (3.86)

where the constants for Eqs. (3.85) and (3.86) are defined in Table 3.6.

In 1998, Buchlin [2] conducted experiments on slender vertical cylinders with diameters between 1.7 and 6.4 mm. The cylinders were 1.8 m long. The wire obtained Joule heating of up to 26,000 W/m^2. It was found that for the 6.4-mm wire, the data were slightly spread, but bounded by the results of Morgan [35] and Carne [4].

In 2004, experiments conducted for water (Pr = 5.5) by Kimura et al. [23] agree closely with the Fujii and Uehara [11] and Isahai et al. [20]. The Isahai et al. correlation is given in [23] and shown below.

$$\frac{Nu_x}{Nu_{x,fp}} = 1 + 0.324 \left(\frac{2x/D}{Nu_{x,fp}} \right)^{0.89} , \quad \frac{x}{D} \leq 0.5 Nu_{x,fp}$$ (3.87)

In 2005, Jarall and Campo [22] experimentally determined local Nusselt numbers for a uniform heat flux boundary in air. The investigators tested three aspect ratios $L/D = 47.5, 68.7$ and 101.3. The local Nusselt number correlation based on the modified Rayleigh number is

$$Nu_x = 1.285 \left(Ra_x^* \frac{x}{D} \right)^{0.165}$$ (3.88)

and is valid for $Ra_x^* \leq 2 \times 10^{12}$. Jarall and Campo compared their correlation to the analytical studies of Mabuchi [31] and Lee et al. [28] and found that their experimental Nusselt number values were approximately 25 % higher than those of the analytical studies.

Table 3.6 Constants for Nagendra et al. [39] correlations

		C_1	C_2	C_3	C_4
Short cylinders	$Ra_D D/L > 10^4$	0.55	0.20	0.60	0.25
Long cylinders	$0.05 < Ra_D D/L < 10^4$	1.33	0.14	1.37	0.16
Wires	$Ra_D D/L < 0.05$	0.90	0.048	0.93	0.05

Fig. 3.1 Local Nusselt number for air ($Pr = 0.708$), water (7.02), and oil ($Pr = 50$) (© Springer 2006), reprinted with permission

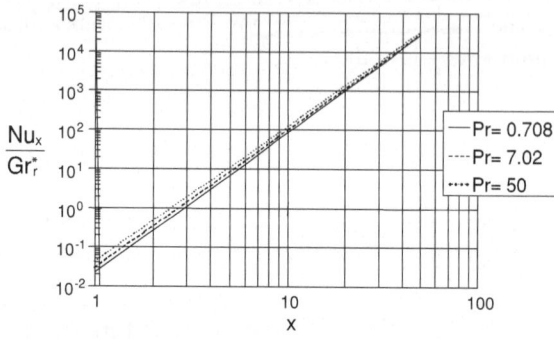

Fig. 3.2 Average Nusselt number for air ($Pr = 0.708$), water (7.02), and oil ($Pr = 50$) (© Springer 2006), reprinted with permission

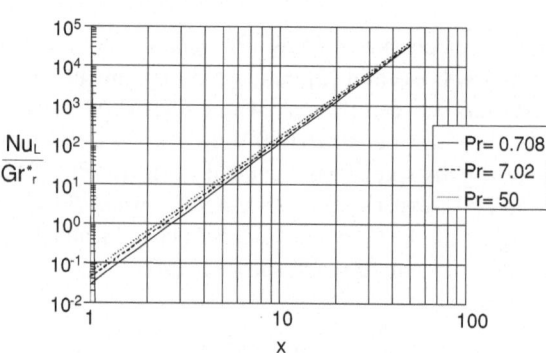

In 2006, Gori et al. [14] carried out an analytical study using the methods of Minkowycz and Sparrow [34] on natural convection around thin vertical cylinders heated at uniform and constant heat flux 0.6 mm $\leq$ D $\leq$ 1.5 mm, $1 \leq \xi \leq 100$, and $0.7 \leq \mathrm{Pr} \leq 730$. The results are displayed in Figs. 3.1 and 3.2. In the figures, the local and average Nusselt numbers were divided by the modified Grashof number Gr_r^* based on the radius of the thin wire.

References

1. Al-Arabi M, Khamis M (1982) Natural convection heat transfer from inclined cylinders. Int J Heat Mass Transfer 25:3–15
2. Buchlin JM (1998) Natural and forced convective heat transfer on slender cylinders. Rev Gen Therm 37:653–660
3. Burmeister L (1993) Convective heat transfer, 2nd edn. Wiley, New York
4. Carne JB (1937) Heat loss by natural convection from vertical cylinders. Lond Edinb Dubl Phil Mag 24:634–653
5. Cebeci T (1974) Laminar-free-convective-heat transfer from the outer surface of a vertical circular cylinder. In: Proceedings of the 5th international heat transfer conference, Tokyo, pp 1–64

6. Churchill SW, Chu HHS (1975) Correlating equations for laminar and turbulent free convection from a vertical plate. Int J Heat Mass Transfer 18:1323–1329

7. Day JC, Zemler MK, Traum MJ, Boetcher SKS (2013) Laminar natural convection from isothermal vertical cylinders: revisiting a classical subject. J Heat Transfer 135:22505-1–9

8. Ede AJ (1967) Advances in free convection. Advances in heat transfer. Academic Press, New York

9. Eigenson L (1940) Les lois gouvernant la transmission de la chaleur aux gaz biatomiques par les parois des cylindres verticaux dans le cas de convection naturelle. Dokl Akad Nauk SSSR 26:440–444

10. Elenbaas W (1948) The dissipation of heat by free convection from vertical and horizontal cylinders. J Appl Phys 19:1148–1154

11. Fujii T, Uehara H (1970) Laminar natural-convective heat transfer from the outside of a vertical cylinder. Int J Heat Mass Transfer 13:607–615

12. Fujii T, Takeuchi M, Fujii M, Suzaki K, Uehara H (1970) Experiments on natural-convection heat transfer from the outer surface of a vertical cylinder to liquids. Int J Heat Mass Transfer 13:753–787

13. Gebhart B, Jaluria Y, Mahajan RL, Sammakia B (1988) Buoyancy-induced flows and transport, Reference edn. Hemisphere Publishing Company, New York

14. Gori F, Serrano MG, Wang Y (2006) Natural convection along a vertical thin cylinder with uniform and constant wall heat flux. Int J Thermophys 27:1527–1538

15. Griffiths E, Davis AH (1922) The transmission of heat by radiation and convection. Tech. rep, Food Investigation Board Special Report No. 9

16. Hanesian D, Kalish RL (1970) Heat transfer by natural convection wiht fluorocarbon gases. IEEE Trans Parts Mater 6:146–148

17. Holman JP (2010) Heat transfer, 10th edn. Mc-Graw-Hill Companies Inc, New York

18. Incropera FP, Dewitt DP, Bergman TL, Lavine AS (2007) Fundamentals of heat and mass transfer, 6th edn. Wiley, Hoboken

19. Incropera FP, Dewitt DP, Bergman TL, Lavine AS (2007) Introduction to heat transfer, 5th edn. Wiley, Hoboken

20. Isahai Y, Suetsugu K, Hattori N (2001) Trans Jpn Soc Mech Eng (Japanese) 67:300–303

21. Jakob M, Linke W (1935) Der Wärmeübergang beim Verdampfen von Flüssigkeiten an senkrechten und waagerechten Flächen. Phys Z 36:267–280

22. Jarall S, Campo A (2005) Experimental study of natural convection from electrically heated vertical cylinders immersed in air. Exp Heat Transfer 18:127–134

23. Kimura F, Tachibana T, Kitamura K, Hosokawa T (2004) Fluid flow and heat transfer of natural convection around heated vertical cylinders (effect of cylinder diameter). JSME Int J Ser B 47:156–161

24. Koch W (1927) Über die Wärmeabgabe geheizter Rohre bei verschiedener Neigung der Rohrachse. Gesundh Ing Beih 1:1–29

25. Kreith F, Manglik R, Bohn M (2011) Principles of heat transfer. Cengage Learning

26. Kyte JR, Madden AJ, Piret EL (1953) Natural-convection heat transfer at reduced pressure. Chem Eng Progr 49:653–662

27. Langmuir I (1912) Convection and conduction of heat in gases. Philos Mag 24:401–422

28. Lee HR, Chen TS, Armaly BF (1988) Natural convection along slender vertical cylinders with variable surface temperature. J Heat Transfer 110:103–108

29. LeFevre EJ (1956) Laminar free convection form a vertical plane surface. In: Proceedings of the ninth international congress of applied mechanics

30. LeFevre EJ, Ede AJ (1956) Laminar free convection from the outer surface of a vertical cylinder. In: Proceedings of the 9th international congress on applied mechanics, Brussels, Belgium, pp 175–183

31. Mabuchi I (1961) Laminar free convection from a vertical cylinder with uniform surface het flux. Trans JSME 27:1306–1313

32. McAdams W (1954) Heat transmission. Mc-Graw-Hill, New York

33. Millsaps K, Pohlhausen K (1958) The laminar free-convective heat transfer from the outer surface of a vertical circular cylinder. J Aeronaut Sci 25:357–360
34. Minkowycz WJ, Sparrow EM (1974) Local nonsimilar solutions for natural convection on a vertical cylinder. J Heat Transfer 96:178–183
35. Morgan V (1975) The overall convective heat transfer from smooth circular cylinders. Adv Heat Transfer 11:199–264
36. Mueller AC (1942) Heat transfer from wires to air in parallel flow. Trans Am Inst Chem Eng 38:613–627
37. Nagendra HR, Tirunarayanan MA, Ramachandran A (1969) Free convection heat transfer from vertical cylinders and wires. Chem Eng Sci 24:1491–1495
38. Nagendra HR, Tirunarayanan MA, Ramachandran A (1970) Free convection heat transfer from vertical cylinders part i: power law surface temperature variation. Nucl Eng Des 16:153–162
39. Nagendra HR, Tirunarayanan MA, Ramachandran A (1970) Laminar free convection from vertical cylinders with uniform heat flux. ASME J Heat Trans 92:191–194
40. Nagendra HR, Tirunarayanan MA, Ramachandran A (1971) Free convection heat transfer from vertical cylinders part ii: exponential surface temperature variation. Nucl Eng Des 16:163–168
41. Ostrach S (1953) An analysis of laminar free convection flow and heat transfer about a flat plate parallel to the direction of the generating body forces. NACA, Report 1111
42. Popiel CO (2008) Free convection heat transfer from vertical slender cylinders: a review. Heat Transfer Eng 29:521–536
43. Popiel CO, Wojtkowiak J, Bober K (2007) Laminar free convective heat transfer from isothermal vertical slender cylinders. Exp Therm Fluid Sci 32:607–613
44. Saunders OA (1936) The effect of pressure upon natural convection in air. Proc Roy Soc A157:278–291 (London)
45. Senftleben H (1951) Die Wärmeabgabe von Körpern verschiedener Form in Flüssigkeiten und Gasen bei freier Strömung. Z Angew Phys 3:361–373
46. Sparrow EM, Gregg JL (1956) Laminar-free-convection heat transfer from the outer surface of a vertical circular cylinder. Trans ASME 78:1823–1829
47. Touloukian YS, Hawkins GA, Jakob M (1948) Heat transfer by free convection from heated vertical surfaces to liquids. T ASME 70:13–18
48. Weise R (1935) Wärmeübergang durch freie Konvecktion an quadratischen platten. Forsch Gebiete Ingenieurw 6:281–292

Chapter 4
Natural Convection Heat Transfer From Inclined Cylinders

Abstract There exists minimal work on natural convection from inclined cylinders in the literature. This is because natural convection from inclined cylinders is a three-dimensional problem as opposed to horizontal and vertical cylinders, which is a two-dimensional problem. The following chapter is organized into inclined cylinders with a temperature boundary condition, heat flux boundary condition, and heat transfer correlations via mass transfer experiments.

Keywords Natural convection · Inclined cylinder · Experimental · Computational · Analytical · Nusselt numbers

4.1 Introduction

Not as much work on natural convection from inclined cylinders is found in the literature. This is mainly because natural convection from horizontal and vertical cylinders is a two-dimensional analytical and numerical problem and natural convection from inclined cylinders is a three-dimensional problem, which makes the flow more complex. As the angle of inclination from the horizontal ϕ increases, the rate of heat transfer decreases from that of the horizontal cylinder. According to Morgan [9], for angles of inclination between 0° and 45°, the heat transfer decreases approximately 8 % from the horizontal cylinder. However, as the angle of inclination approaches the vertical cylinder, the heat transfer rate begins to decrease more rapidly.

Further, for larger Rayleigh numbers, the average Nusselt number of a cylinder inclined from the horizontal at angle ϕ can be correlated by replacing D in the equations for the average Nusselt number from horizontal cylinders with $D/\cos \phi$ [3].

© The Author(s) 2014
S.K.S. Boetcher, *Natural Convection from Circular Cylinders*, SpringerBriefs
in Thermal Engineering and Applied Science, DOI: 10.1007/978-3-319-08132-8_4

4.2 Temperature Boundary Conditions

In 1976, Oosthuizen [10, 11] numerically and experimentally determined the natural convection from isothermal inclined cylinders for $8.00 \leq L/D \leq 16.00$ and $40,000 \leq Gr_D \leq 90,000$ and $0 \leq \phi \leq 90$, where ϕ is the angle of inclination measured from the horizontal. For $Pr = 0.7$, the average diameter-based Nusselt number was found to be

$$\frac{Nu_D}{(Gr_D \cos \phi)^{0.25}} = 0.42 \left[1 + \left(\frac{1.31}{\left(\frac{L}{D \tan \phi} \right)^{0.25}} \right)^8 \right]^{0.125} \tag{4.1}$$

In 1978, Raithby and Holland [12] used an approximate analytical method to determine the average heat transfer coefficients for isothermal inclined cylinders. The average Nusselt number corresponding to Raithby and Holland's thin-layer analysis is

$$Nu_D = \left[0.772 + \frac{0.228}{1.0 + 0.676 p^{1.23}} \right] [\cos \phi + (D/L) \sin \phi]^{1/4} C_l Ra_D^{1/4} \tag{4.2}$$

where

$$C_l = 0.671 \left[1 + \left(\frac{0.492}{Pr} \right)^{9/16} \right]^{4/9} \tag{4.3}$$

and

$$p = (2L/D) \cot \phi \tag{4.4}$$

In 1981, Stewart [14] experimentally determined natural convection from inclined cylinders for $6 \leq L/D \leq 12$. The following average Nusselt number correlation was developed

$$\frac{Nu_D}{(Ra_D \cos \phi)^{0.25}} = 0.53 + 0.555 \left[\left(\frac{D}{L \cos \phi} \right)^{0.25} - \left(\frac{D}{L} \right)^{0.25} \right] \tag{4.5}$$

According to Stewart, the correlation fits between the limits of the horizontal cylinder correlation which was proposed by McAdams [8] and the vertical cylinder correlation proposed by Jakob and Linke [5].

In 1982, Al-Arabi and Khamis [1] experimentally determined average Nusselt numbers for air ($Pr = 0.7$). The isothermal boundary condition was employed using steam condensation. These investigators used the angle θ, which is the angle of inclination of the cylinder as measured from the vertical position.

For laminar flow and $1.08 \times 10^4 \leq Gr_D \leq 6.9 \times 10^5$ and $9.88 \times 10^7 \leq Gr_L Pr \leq (Gr_L Pr)_{Cr}$ where

$$(Gr_L Pr)_{Cr} = 2.6 \times 10^9 + 1.1 \times 10^9 \tan\theta \qquad (4.6)$$

the average Nusselt number is

$$Nu_L = \left[2.9 - 2.32(\sin\theta)^{0.8}\right] Gr_D^{-1/12}(Gr_L Pr)^{1/4+1/12(\sin\theta)^{1.2}} \qquad (4.7)$$

and for turbulent flow when $1.08 \times 10^4 \leq Gr_D \leq 6.9 \times 10^5$ and $(Gr_L Pr)_{Cr} \leq Gr_L Pr \leq 2.95 \times 10^{10}$

$$Nu_L = \left[0.47 + 0.11(\sin\theta)^{0.8}\right] Gr_D^{-1/12}(Gr_L Pr)^{1/3} \qquad (4.8)$$

Al-Arabi and Khamis also experimentally determined local Nusselt numbers for air.

For laminar flow and $1.08 \times 10^4 \leq Gr_D \leq 6.9 \times 10^5$ and $1.63 \times 10^8 \leq Gr_x Pr \leq (Gr_x Pr)_{Cr-1}$, the local Nusselt number is

$$Nu_x = \left[2.3 - 1.72(\sin\theta)^{0.8}\right] Gr_D^{-1/12}(Gr_x Pr)^{1/4+1/12(\sin\theta)^{1.2}} \qquad (4.9)$$

and for turbulent flow when $1.08 \times 10^4 \leq Gr_D \leq 6.9 \times 10^5$ and $(Gr_x Pr)_{Cr-2} \leq Gr_x Pr \leq 2.3 \times 10^{10}$

$$Nu_x = \left[0.42 + 0.16(\sin\theta)^{0.8}\right] Gr_D^{-1/12}(Gr_x Pr)^{1/3} \qquad (4.10)$$

where $(Gr_x Pr)_{Cr-1}$ represents the critical Rayleigh number at which the laminar region ends

$$(Gr_x Pr)_{Cr-1} = 2.16 \times 10^9 + 0.286 \times 10^9 \tan\theta \qquad (4.11)$$

and $(Gr_x Pr)_{Cr-2}$ represents the critical Rayleigh number where the turbulent region begins

$$(Gr_x Pr)_{Cr-2} = 4.4 \times 10^9 \qquad (4.12)$$

In 1992, Li and Tarasuk [7] performed an experimental study to determine the average heat transfer coefficients for inclined isothermal cylinders with aspect ratios of 14 and 21 and for angles of inclination ϕ of $0°$, $45°$, $60°$, $75°$, and $90°$. A fourth-order polynomial correlation for the average Nusselt number is given as

Table 4.1 Table of coefficients for Eqs. (4.13–4.15)

$A_1 = 0.5925$	$B_1 = 0.2295$
$A_2 = 0.2278 \times 10^{-2}$	$B_2 = 0.1553 \times 10^{-2}$
$A_3 = -0.1436 \times 10^{-3}$	$B_3 = -0.7396 \times 10^{-4}$
$A_4 = 0.1877 \times 10^{-5}$	$B_4 = 0.1157 \times 10^{-5}$
$A_5 = 0.9860 \times 10^{-8}$	$B_5 = -0.5783 \times 10^{-8}$

$$Nu_D = m(\phi) Ra_D^{n(\phi)} \tag{4.13}$$

where

$$m(\phi) = A_1 + A_2\phi + A_3\phi^2 + A_4\phi^3 + A_5\phi^4 \tag{4.14}$$

and

$$n(\phi) = B_1 + A_2\phi + B_3\phi^2 + B_4\phi^3 + B_5\phi^4 \tag{4.15}$$

The coefficients A and B are defined in Table 4.1.

In 2009, Kalendar and Oosthuizen [6] studied the natural convection from inclined isothermal cylinders with heated tops. The total Nusselt number including the sides and heated top based on the height of the cylinder is

$$\frac{Nu_L}{Ra_L^{0.284+0.005\sin\theta}} = 0.2 + \frac{0.63}{((D/L)Ra^{0.25})^{0.59}} \tag{4.16}$$

Here, θ is the angle of inclination from the vertical.

4.3 Heat Flux Boundary Conditions

In 1980, Al-Arabi and Salman [2] experimentally determined local and average Nusselt numbers from an inclined cylinder of $L/D = 25$ in air subjected to constant heat flux boundary conditions and angles of inclination θ from the vertical of 30°, 45°, 60°, and 90°.

An expression for the local heat transfer coefficient was found to be

$$Nu_x = 0.545 - 0.387(\sin\theta)^{1.462}(Gr_x Pr)^{1/4+1/12(\sin\theta)^{1.75}} \tag{4.17}$$

The correlation for the average heat transfer coefficient is

$$Nu_L = 0.60 - 0.488(\sin\theta)^{1.03}(Gr_L Pr)^{1/4+1/12(\sin\theta)^{1.75}} \tag{4.18}$$

In 1986, Fujii et al. [3] performed natural convection experiments for thin wires 50.4 μm in diameter and 270 mm long for $0° \leq \phi \leq 85°$ and $10^{-4} \leq Gr_D Pr \leq 10^{-3}$

in air ($Pr = 0.7$). The results of the study have been correlated in the following equation

$$\frac{2}{Nu_D} = \ln\left[1 + \frac{3.3}{C(Pr)(Gr_D Pr \cos\phi)^n}\right] \tag{4.19}$$

where

$$n = 0.25 + \frac{1}{10 + 5(Gr_D Pr)^{0.175}} \tag{4.20}$$

and

$$C(Pr) = \frac{0.671}{\left[1 + (0.492/Pr)^{9/16}\right]^{4/9}} \tag{4.21}$$

4.4 Heat Transfer Correlations via Mass Transfer Experiments

In 1982, Sedahmed and Shemilt [13] performed mass transfer experiments to determine average Nusselt numbers for inclined cylinders. In the experiments, $4.65 \leq L/D \leq 14.3$, $1.9 \times 10^{10} \leq Ra_L \cos\phi \leq 3.8 \times 10^{11}$, and $Pr = 2{,}300$. The resulting analogous average Nusselt number correlation based on the length of the cylinder is

$$Nu_L = 0.498(Ra_L \cos\phi)^{0.28} \tag{4.22}$$

In 2012, Heo and Chung [4] also performed mass transfer experiments using a copper electroplating system. The ranges used in the experiments were $L/D = 3.7 - 25$, $Ra_D = 1.6 \times 10^8 - 5.07 \times 10^{10}$, $Ra_L = 2.64 \times 10^{12} - 1.54 \times 10^{13}$, and $Pr = 2{,}094$.

The following correlations were arrived at for laminar flow

$$Nu_D = 0.3 Ra_D^{0.25}(1 + 0.7\cos\phi) \tag{4.23}$$

$$Nu_L = 0.67 Ra_L^{0.25}(1 + 1.44 Ra_D^{-0.04}\cos\phi) \tag{4.24}$$

and for turbulent flow

$$Nu_D = 0.13 Ra_D^{0.3}(1 + 0.6\cos\phi) \tag{4.25}$$

$$Nu_L = 0.26 Ra_L^{0.28}(1 + 1.89 Ra_D^{-0.044}\cos\phi) \tag{4.26}$$

References

1. Al-Arabi M, Khamis M (1982) Natural convection heat transfer from inclined cylinders. Int J Heat Mass Transfer 25:3–15
2. Al-Arabi M, Salman Y (1980) Laminar natural convection heat transfer from an inclined cylinder. Int J Heat Mass Transfer 23:45–51
3. Fujii T, Koyama S, Fujii M (1986) Experimental study of free convection heat transfer from an inclined fine wire to air. In: Heat transfer 1986, proceedings of the eight international heat transfer conference, San Francisco
4. Heo J, Chung B (2012) Natural convection heat transfer on the outer surface of inclined cylinders. Chem Eng Sci 73:366–372
5. Jakob M, Linke W (1935) Der Wärmeübergang beim Verdampfen von Flüssigkeiten an senkrechten und waagerechten Flächen. Phys Z 36:267–280
6. Kalendar A, Oosthuizen P (2009) Natural convection heat transfer from an inclined isothermal cylinder with an exposed top surface mounted on a flat adiabatic base. In: Proceedings of the ASME 2009 international mechanical engineering congress and exposition
7. Li J, Tarasuk J (1992) Local free convection around inclined cylinders in air: an interferometric study. Exp Therm Fluid Sci 1992:235–242
8. McAdams W (1954) Heat transmission. Mc-Graw-Hill, New York
9. Morgan V (1975) The overall convective heat transfer from smooth circular cylinders. Adv Heat Transfer 11:199–264
10. Oosthuizen P (1976) Experimental study of free convective heat transfer from inclined cylinders. J Heat Transfer 98:672–674
11. Oosthuizen P (1976) Numerical study of some three-dimensional laminar free convective flows. J Heat Transfer 98:570–575
12. Raithby G, Hollands K (1978) Analysis of heat transfer by natural convection (or film condensation) for three dimensional flows. In: Heat transfer 1978, proceedings of the sixth international heat transfer conference, Toronto
13. Sedahmed G, Shemilt L (1982) Natural convection mass transfer at cylinders in different positions. Chem Eng Sci 37:159–166
14. Stewart W (1981) Experimental free convection from an inclined cylinder. J Heat Transfer 103:817–819